Exploring Algebra 1

with THE GEOMETER'S
SKETCHPAD®
VERSION 5

Addressing the Common Core State Standards for Mathematics

Key Curriculum Press
INNOVATORS IN MATHEMATICS EDUCATION

Writers:	Paul Kunkel, Steven Chanan, Scott Steketee
Contributing Writers:	Allan Bergmann Jensen, Ralph Pantozzi, Jennifer North Morris, Brooke Precil, Daniel Scher, Nathalie Sinclair, Dan Bennett, Eric Bergofsky, Eric Kamischke
Reviewers:	Molly Jones, Dan Lufkin, Marsha Sanders-Leigh, John Threlkeld, Pat Brewster
Editors:	Scott Steketee, Andres Marti, Elizabeth DeCarli
Contributing Editors:	Ladie Malek, Josephine Noah, Cindy Clements, Sharon Taylor
Editorial Assistant:	Tamar Chestnut
Production Director:	Christine Osborne
Production Editors:	Angela Chen, Andrew Jones, Christine Osborne
Other Contributers:	Judy Anderson, Jason Luz, Aaron Madrigal, Marilyn Perry
Copyeditors:	Tom Briggs, Jill Pellarin
Cover Designer:	Suzanne Anderson
Cover Photo Credit:	Ocean Photography
Printer:	Lightning Source, Inc.
Executive Editor:	Josephine Noah
Publisher:	Steven Rasmussen

Key Curriculum Press
1150 65th Street
Emeryville, CA 94608
510-595-7000
editorial@keypress.com
www.keypress.com

ISBN: 978-1-60440-221-6
10 9 8 7 6 5 4 3 2 1   DOH   15 14 13 12

Contents

Chapter 4: Solving Equations and Inequalities

Chapter 5: Coordinates, Slope, and Distance

Chapter 6: Variations and Linear Equations

Chapter 7: Quadratic Equations

Exploring Algebra 1 with The Geometer's Sketchpad
© 2012 Key Curriculum Press

Downloading Sketchpad Documents

Getting Started

All Sketchpad documents (sketches) for *Exploring Algebra 1 with The Geometer's Sketchpad* are available online for download.

- Go to www.keypress.com/gsp5modules.
- Log in using your Key Online account, or create a new account and log in.
- Enter this access code: EA1HS864
- A Download Files button will appear. Click to download a compressed (.zip) folder of all sketches for this book.

The downloadable folder contains all of the sketches you need for this book, organized by chapter and activity. The sketches require The Geometer's Sketchpad Version 5 software to open. Go to www.keypress.com/gsp/order to purchase Sketchpad, or download a trial version from www.keypress.com/gsp/download.

Types of Sketches

Student Sketches: In most activities, students use a prepared sketch that provides a model, a simulation, or a complicated construction to investigate relationships. The name of a Student Sketch usually matches the activity title and is referenced on the Student Worksheet for the activity, such as **Binomial Product.gsp.**

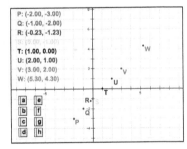

Presentation Sketches: Some activities include sketches designed for use with a projector or interactive whiteboard, either for a teacher presentation or a whole-class activity. Presentation Sketches often have action buttons to enhance your presentation. For activities in which students create their own constructions, the Presentation Sketch can be used to speed up, summarize, or review the mathematical ideas from the activity. Many activities include a separate page of Presenter Notes to help you facilitate a whole-class presentation. The name of a Presentation Sketch always ends with the word "Present," such as **Points Line Up Present.gsp.**

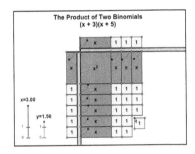

Sketchpad Resources

Sketchpad Learning Center

The Learning Center provides a variety of resources to help you learn how to use Sketchpad, including overview and classroom videos, tutorials, Sketchpad Tips, sample activities, and links to online resources. You can access the Learning Center through Sketchpad's start-up screen or through the Help menu.

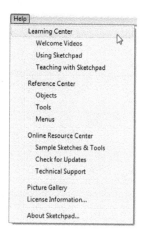

The Learning Center has three main sections:

Welcome Videos

These videos introduce Sketchpad from the point of view of students and teachers, and give an overview of the big ideas and new features of Sketchpad 5.

Using Sketchpad

This section includes 12 self-guided tutorials with embedded videos, 70 Sketchpad Tips, and links to local and online resources.

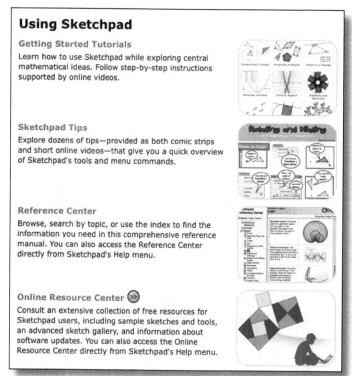

Teaching with Sketchpad

This section includes videos and articles describing how teachers make effective use of Sketchpad and how it affects their students' attitudes and mathematical understanding. There are over 40 sample activities, each with an overview, teaching notes, student worksheet, and sketches, that you can use with students to support your subject area, level, and curriculum.

Other Sketchpad Resources

Exploring Algebra 1 with The Geometer's Sketchpad activities are an excellent introduction to Sketchpad for both students and teachers. If you want to learn more, Sketchpad contains resources for beginning and advanced users.

- **Reference Center:** This digital resource, which is accessed through the Help menu, is the complete reference manual for Sketchpad, with detailed information on every object, tool, and menu command. The Reference Center includes a number of How-To sections, an index, and full-text search capability.

- **Online Resource Center:** The Geometer's Sketchpad Resource Center (www.dynamicgeometry.com) contains many sample sketches and advanced toolkits, links to other Sketchpad sites, technical information (including updates and frequently asked questions), and detailed documentation for JavaSketchpad, which allows you to embed dynamic constructions in a web page.

- **Sketch Exchange:** The Sketchpad Sketch Exchange™ (sketchexchange.keypress.com) is a community site where teachers share sketches and other resources with Sketchpad users. Browse by keyword or topic for sketches that interest you, or ask questions and share ideas in the forum.

- **Sample Sketches & Tools:** You can access many sketches, including some with custom tools, through Sketchpad's Help menu. You can use some sample sketches as demonstrations, others to get tips and information about particular constructions, and others to access custom tools that you can use to perform special constructions. These sketches are also available under General Resources at the Sketchpad Resource Center (www.dynamicgeometry.com).

- **Sketchpad LessonLink™:** This online subscription service includes a library of more than 500 prepared activities (including those used in this book) aligned to leading math textbooks and state standards for grades 3–12. For more information, additional sample activities, or a trial subscription, go to www.keypress.com/sll.

- **Online Courses:** Key Curriculum Press offers moderated online courses that last six weeks, allowing you to immerse yourself in learning how to use Sketchpad in your teaching. For more information, see Sketchpad's Learning Center, or go to www.keypress.com/onlinecourses.

- **Other Professional Development:** Key Curriculum Press offers free webinars on a regular basis. You can also arrange for one-day or three-day face-to-face workshops for your district or school. For more information, go to www.keypress.com/pd.

Addressing the Common Core State Standards for Mathematics

The Common Core State Standards emphasize the importance of students gaining expertise with a variety of mathematical tools, including dynamic geometry® software such as The Geometer's Sketchpad. The Standards for Mathematical Practice recognize that effective use of technology is one aspect of mathematical proficiency.

5. Use appropriate tools strategically.

Mathematically proficient students consider the available tools when solving a mathematical problem. These tools might include pencil and paper, concrete models, a ruler, a protractor, a calculator, a spreadsheet, a computer algebra system, a statistical package, or dynamic geometry software. Proficient students are sufficiently familiar with tools appropriate for their grade or course to make sound decisions about when each of these tools might be helpful, recognizing both the insight to be gained and their limitations. They are able to use technological tools to explore and deepen their understanding of concepts. (Common Core State Standards for Mathematics, 2010, www.corestandards.org)

This collection of Sketchpad activities uses the strengths of dynamic mathematical representations to enrich the study of first-year algebra. In the first two chapters, students use dynamic models to review and deepen their understanding of foundational concepts for algebra, such as integer operations, ratio and proportion, and exponent properties. Then students explore equivalent expressions, and solve equations, inequalities, and systems of equations. Students explore linear and quadratic functions through multiple dynamic representations, including tables, graphs, and equations.

The following content clusters from the Common Core State Standards for Mathematics for Grade 8 and High School are addressed in this collection.

Grade 8

Expressions and Equations

- Work with radicals and integer exponents.
- Understand the connections between proportional relationships, lines, and linear equations.
- Analyze and solve linear equations and pairs of simultaneous linear equations.

Functions

- Define, evaluate, and compare functions.
- Use functions to model relationships between quantities.

High School: Algebra

Seeing Structure in Expressions

- Interpret the structure of expressions
- Write expressions in equivalent forms to solve problems

Arithmetic with Polynomials and Rational Functions

- Perform arithmetic operations on polynomials
- Understand the relationship between zeros and factors of polynomials

Creating Equations

- Create equations that describe numbers or relationships

Reasoning with Equations and Inequalities

- Understand solving equations as a process of reasoning and explain the reasoning
- Solve equations and inequalities in one variable
- Solve systems of equations
- Represent and solve equations and inequalities graphically

High School: Functions

Interpreting Functions

- Understand the concept of a function and use function notation
- Interpret functions that arise in applications in terms of the context
- Analyze functions using different representations

Building Functions

- Build a function that models a relationship between two quantities
- Build new functions from existing functions

Linear, Quadratic, and Exponential Models

- Construct and compare linear and exponential models and solve problems
- Interpret expressions for functions in terms of the situation they model

Fundamental Operations

1

Adding Integers

Students use an animated model to add integers on a number line. They investigate addition of two positive numbers, two negative numbers, and a positive and a negative number.

Subtracting Integers

Students use an animated model for subtracting integers on the number line, and see the second number being flipped before it's added to the first number. Students investigate subtraction of positive numbers and subtraction involving negative numbers.

Raz's Magic Multiplying Machine

Students explore multiplication dynamically, dragging the input values and observing the behavior of the output value. As they observe the dynamic behavior of the values, students discover the special roles that zero and one play in multiplication and the rules regarding the sign of a product.

Multiple Models of Multiplication

Students work with four different models of multiplication and use each model to solve problems and investigate properties of multiplication. Students compare the four models, particularly with regard to how they make sense of negative operands.

Mystery Machines

Students explore several mystery machines built on number lines that are missing their numbers. By dragging and observing markers representing a multiplication problem or an addition problem, students determine the locations of hidden numbers. Students manipulate input markers on other machines to determine the arithmetic operation performed by the machine.

Dividing Real Numbers

Students run a multiplication machine in reverse to solve division problems, to observe the relationship between multiplication and division, and to investigate properties of division.

The Commutative Property

Students use dynamic models of the four arithmetic operations to determine which of the operations are commutative.

The Associative Property

Students build and investigate a dynamic model in order to determine which arithmetic operations are associative.

Identity Elements and Inverse

Students build and manipulate a dynamic model for each arithmetic operation in order to determine whether the operation has an identity element, and if so, what that identity element is. For those operations with identity elements, students use another model to find the inverses of particular values, and describe what pattern connects a number to its inverse.

Exploring Properties of Operations

Students use arithmetic machines to explore various properties of the four fundamental arithmetic operations. They use each machine to perform an operation on two variables, and draw conclusions about the similarities and differences between the operations.

 ACTIVITY NOTES

Use this activity as an introduction to integer addition for pre-algebra students, as a start-of-the-year refresher for Algebra 1 students, or as a supplemental activity for any student having difficulty with the topic. It's important for students to have a mental image of operations on integers. Even strong students who rely on verbal rules make careless mistakes that could be avoided by having an internalized picture.

The picture of addition presented here is a geometric model in which each number is represented by a vector. (The activity calls them *arrows* because students may not be familiar with the term *vector*.) Vectors incorporate both magnitude and direction (representing the absolute value and the sign of the integer), so practice with this model helps students understand how the signs of the addends come into play.

This activity contains lots of questions for students, who develop their understanding through the process of manipulating the sketch and describing what they observe. Encourage them to write clear and detailed explanations (and to use complete sentences) when they answer the questions; the extra time it takes them to do so is well spent.

If there's time and you have a presentation computer with a projector, have different students use Sketchpad to demonstrate to the class their observations or the problems they made up. It's a big help to students if they can listen to, evaluate, and discuss the descriptions and conclusions of their classmates.

INVESTIGATE

Students may be unfamiliar with *model* as a transitive verb; consider reviewing with them the various uses of this word.

Q1 In their final positions, the second arrow starts from where the first arrow ends, and the answer (13) is at the end of the second arrow. Encourage students to be detailed and specific in their answer to this question.

Q2 Answers will vary but should include only positive numbers.

Q3 Each lower arrow is exactly the same size and direction as the corresponding upper arrow.

Q4 The sum of $-6 + (-3)$ is -9.

Q5 Answers will vary but should include only negative numbers.

Q6 Whether adding two negative or two positive numbers, both arrows go the same way, taking the sum farther away from the center of the number line (farther away from zero). The difference is that the arrows go to the right when the numbers are positive but go to the left when they're negative.

Q7 When you add two negative numbers, you cannot get a positive sum. Both numbers take the sum in the negative direction from zero, so the sum must be negative.

Q8 As students model various problems, walk around the room and observe them to make sure they can model any problem they are given.

$$7 + (-4) = 3 \qquad -4 + 7 = 3$$
$$-6 + 2 = -4 \qquad 2 + (-6) = -4$$
$$-3 + 7 = 4 \qquad 3 + (-7) = -4$$
$$2 + (-5) = -3 \qquad -2 + 5 = 3$$

Q9 When you add a positive and a negative integer, the number that has the larger absolute value tells you whether the answer will be positive or negative. In other words, the sign of the result is the same as the sign of the longer arrow.

EXPLORE MORE

Q10 Each student will model different problems. In every case, the two numbers must be opposites, so that their arrows are the same length but point in opposite directions.

Q11 The order does not matter when you add two numbers. The arrows determine how far you go and in which direction, and it doesn't matter if you follow the first arrow and then the second, or if you follow the second arrow and then the first.

WHOLE-CLASS PRESENTATION

Start the whole-class presentation by animating the addition of two positive integers (Q1–Q3 of the activity). Open the sketch **Adding Integers Present.gsp** and press the step-by-step buttons one at a time, pausing between animations. Ask students to describe what they see as the animation

progresses, and be sure to get observations from several different students. Press the *Reset* button, change the problem by dragging both circles (while leaving the numbers positive), and press the step-by-step buttons again.

Next animate the addition of two negative numbers (Q4–Q7 of the activity). Press *Reset,* drag the numbers so they are both negative, and ask students to predict what will happen now. Use the step-by-step buttons to test their conjectures. Without resetting, ask questions Q6 and Q7, and experiment by dragging to change the values of the numbers.

When students are satisfied with the results of adding two negative numbers, animate the addition of numbers with different signs. Reset again and drag the circles so one of the numbers is positive and one is negative. Ask students to predict how the arrows will behave. (Try to get students to concentrate on the behavior of the model rather than on the numeric answer.) Use the buttons again to show the behavior. Model several more problems (such as those in Q8) involving a positive and a negative number.

Finish the class discussion using Q9, Q10, and Q11. When students propose an answer to one of these questions, have them manipulate the sketch to show why their answer makes sense.

Adding Integers

In this activity you'll add integers using an animated Sketchpad model.

INVESTIGATE

1. Open **Adding Integers.gsp.** This sketch models the addition problem 8 + 5.

2. Press the *Present All* button to see the model in action.

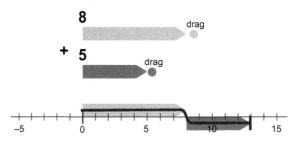

Q1 How does the final position of the arrows show the answer for 8 + 5?

3. Press the *Reset* button, and then drag the circles to model 2 + 6.

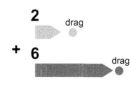

4. This time, show the animation step by step: Press the *Show Steps* button, and then press each numbered button in order.

For each problem, press the buttons to show the result.

Q2 Drag the circles and press the buttons to model two other addition problems using only positive integers. Record each problem and the result.

Q3 How do the two upper arrows in the sketch relate to the two lower arrows?

Q4 Model −6 + (−3). What's the sum?

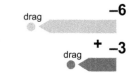

Q5 Model two more addition problems using negative integers. Record each problem and its result.

Q6 How is adding two negative numbers similar to adding two positive numbers? How is it different?

Q7 Can you add two negative numbers and get a positive sum? Explain.

Exploring Algebra 1 with The Geometer's Sketchpad
© 2012 Key Curriculum Press

Q8 Model the following eight problems. Record each problem and its answer.

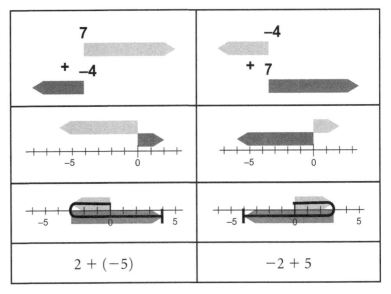

| $2 + (-5)$ | $-2 + 5$ |

Q9 When you add a positive and a negative integer, how can you look at the numbers and tell whether the answer will be positive or negative?

EXPLORE MORE

Q10 Model four problems for which the sum is zero. Make the first number positive in two problems and negative in two problems. Write down the problems you used. What must be true about two numbers if their sum is zero?

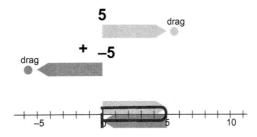

Q11 When you add two numbers, does the order matter? In other words, is $-3 + 5$ the same as $5 + (-3)$? Using the sketch, explain why your answer makes sense.

Subtracting Integers

Use this activity as an introduction to integer subtraction for pre-algebra students, as a start-of-the-year refresher for Algebra 1 students, or as a supplemental activity for any student having difficulty with the topic. It's important for students to have a mental image of operations on integers. Even strong students who rely on verbal rules make careless mistakes that could be avoided by having an internalized picture.

The picture of subtraction presented here is a geometric model in which each number is represented by a vector. (The activity calls them *arrows* because students may not be familiar with the term *vector*.) Vectors incorporate both magnitude and direction (representing the absolute value and the sign of the integer), so practice with this model helps students understand how the signs of the operands come into play.

The questions are critical in encouraging students to internalize the model presented in this activity. Make sure students write clear and detailed explanations (and use complete sentences) when they answer the questions; the extra time it takes them to do so is time well spent.

If there's time and you have a presentation computer with a projector, have different students use Sketchpad to demonstrate to the class their observations or the problems they made up. It's a big help to students if they can listen to, evaluate, and discuss the descriptions and conclusions of their classmates.

INVESTIGATE

These notes sometimes use the terms *minuend* (first number) and *subtrahend* (second number), but these terms are not used in the student material. If you do use them with students, be sure to explain them carefully.

The concept of *additive inverse* is not named, but it plays a prominent role in the animation. You should discuss with the class why the second number must be flipped, even if you don't give a name to that operation.

Q1 During the animation, the arrow for 5 flips from the right to the left. This shows which way the second arrow must go in order to subtract it from the first.

Q2 In their final positions, the flipped second arrow starts from where the first arrow ends, and the answer (3) is at the end of the second arrow. Encourage students to be detailed and specific in their answer to this question.

Exploring Algebra 1 with The Geometer's Sketchpad
© 2012 Key Curriculum Press

Q3 Answers will vary. Students should describe the arrow flipping from right to left; encourage them to explain in their own words why it needs to flip in order to do subtraction.

Q4 Answers will vary but should include only problems in which a positive minuend is smaller than a positive subtrahend.

Q5 If both numbers are positive, the result will be positive if the first number (minuend) is larger, and negative if the second number (subtrahend) is larger.

Q6 Some students will record direct observations, and others will interpret those observations. Typical answers will be similar to the following.

Observation: In this problem, $4 - (-3)$, the second arrow starts out pointing to the left, so when it flips it turns around and points to the right.

Interpretation: The second number starts out negative, so when it flips it becomes positive.

Q7 The problems students create will vary. Because the first number is positive and the second negative, the models have in common that, after flipping, both arrows point to the right, and the result must be positive.

Q8 Problems will vary. Because the first number is negative and the second positive, after flipping, both arrows point to the left, and the result is negative.

Q9 As students model various problems, walk around the room and observe them to make sure they can model any problem they are given.

$$7 - (-4) = 11 \qquad\qquad -4 - 7 = -11$$
$$-6 - (-2) = -4 \qquad\qquad -3 - (-6) = 3$$
$$-3 - 8 = -11 \qquad\qquad -3 - (-8) = 5$$
$$2 - (-7) = 9 \qquad\qquad -2 - 7 = -9$$

Q10 Written as addition problems, these problems become

$$7 + 4 = 11 \qquad\qquad -4 + (-7) = -11$$
$$-6 + 2 = -4 \qquad\qquad -3 + 6 = 3$$
$$-3 + (-8) = -11 \qquad\qquad -3 + 8 = 5$$
$$2 + 7 = 9 \qquad\qquad -2 + (-7) = -9$$

In each case, to subtract you can change the sign of the second number and add them. This is similar to the way the second arrow flips before the animation shows the answer.

EXPLORE MORE

Q11 For a subtraction problem to have an answer of zero, the two numbers being subtracted must be the same.

Q12 To make the difference the same as the first number, the second number must be zero.

Q13 To make the difference the same as the second number, the first number must be twice as big as the second. For instance, $6 - 3 = 3$, and $-8 - (-4) = -4$.

Q14 The order does matter when you subtract numbers, because only the second arrow is flipped. More sophisticated students will observe that the order matters only if the second number is nonzero, because flipping zero has no effect.

WHOLE-CLASS PRESENTATION

Start the whole-class presentation by animating the subtraction of two positive integers (Q1–Q5 of the activity). Open the sketch **Subtracting Integers Present.gsp** and press the step-by-step buttons one at a time, pausing between animations. Ask students to describe what they see as the animation progresses, and be sure to get observations from several different students. Press the *Reset* button, change the problem by dragging both circles (while leaving the numbers positive), and press the step-by-step buttons again. Pay special attention to Q3 and Q5.

Next animate subtraction problems in which the first number is positive and the second number is negative (Q6–Q7 of the activity). Press *Reset*, make the first number positive and the second negative, and ask students to predict what will happen now. Test their conjectures using the step-by-step buttons. Repeat for several more problems.

Animate subtraction problems like those in Q8 and Q9, and record the answers for each of the problems in Q9. Ask students what patterns they see, and how they could predict the answer from the two numbers being subtracted.

For Q10, ask students to make an addition problem for each of the problems from Q9, and test their addition problems using page 2 of the sketch. Switching back and forth between page 1 and page 2 will reinforce for students the idea of using addition to rewrite a subtraction problem.

Continue the class discussion with as many of the Explore More questions (Q11–Q14) as are appropriate for the class and the available time.

Finish by having students summarize in their own words the relationship between subtraction and addition.

Subtracting Integers

In this activity you'll subtract integers using an animated Sketchpad model.

INVESTIGATE

1. Open **Subtracting Integers.gsp.** The sketch models the subtraction problem $8 - 5$.

2. Press the *Present All* button to see the model in action.

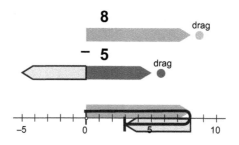

Q1 During the animation, what happens to the arrow for 5?

Q2 How does the final position of the bottom arrows show the answer for this subtraction problem?

3. Press the *Reset* button, and then drag the circles to model $2 - 6$.

4. This time, show the animation step by step: Press the *Show Steps* button, and then press each numbered button in order.

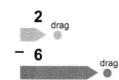

Q3 Describe in your own words what the *3. Make Inverse* step does.

For each problem, press the buttons to show the result.

Q4 Drag the circles to model two more subtraction problems that use positive integers but have a negative result. Record each problem and its result.

Q5 If both numbers in a subtraction problem are positive, how can you tell if the answer will be positive or negative?

Q6 Model $4 - (-3)$. What's different about the *3. Make Inverse* step this time?

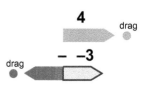

Q7 Model two more problems in which the first number is positive and the second number is negative. Record each problem. What do these models have in common?

Q8 Model three problems in which the first number is negative and the second number is positive. Record each problem. What do these models have in common?

Exploring Algebra 1 with The Geometer's Sketchpad
© 2012 Key Curriculum Press

Q9 Model the following eight problems. Record each problem and its answer.

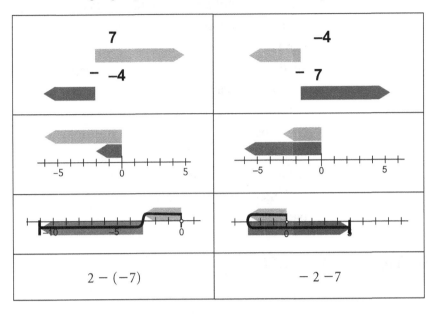

Q10 For each subtraction problem above, write an addition problem that has the same first number and the same answer. What do you notice?

For instance,
7 − (−4) = 11,
so fill in the blank:
7 + ___ = 11.

EXPLORE MORE

Q11 Model four subtraction problems for which the difference is zero. Make the first number positive in two problems and negative in two problems. Write down the problems you used. What must be true about two numbers if their difference is zero?

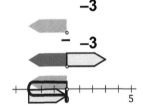

Q12 Model four subtraction problems in which the difference is the same as the first number. What must be true of these problems?

Q13 Model four subtraction problems in which the difference is the same as the second number. What must be true of these problems?

Q14 When you subtract two numbers, does the order matter? In other words, is −3 − (−5) the same as −5 − (−3)? Explain in terms of the model why your answer makes sense.

Raz's Magic Multiplying Machine

This is a useful activity, both for reviewing multiplication of positive and negative numbers and for developing number sense in general. Students can use this simple machine to explore many important number properties, such as multiplication of negatives, square roots, ratios and proportions, and multiplication by numbers between 0 and 1 (and also between -1 and 0). The key is for students to explain mathematically the behavior they observe: "What is it about multiplication that makes the machine behave like this?" Encourage students to answer the questions with detailed descriptions that include both observations and explanations, and encourage them to do their own explorations and observe their own patterns.

STATIC MULTIPLICATION

Q1 $0.5 \cdot 6 = 3$.

Q2 Answers will vary. In a class discussion, you might ask students what different *types* of answers to this question are possible. (One possibility is that both factors could be positive integers. Another is that one could be a positive integer and the other could be a decimal between 0 and 1.)

Q3 The only two possible locations are at 0 $(0 \cdot 0 = 0)$ and 1 $(1 \cdot 1 = 1)$. (To see that these are the only two solutions, solve the equation $x \cdot x = x$.)

Q4 a must be at -1.

Q5 No, all three markers cannot be to the left of 0. When a and b are to the left of 0, $a \cdot b$ is to its right, since a negative times a negative equals a positive. When $a \cdot b$ is to the left of 0, one of the factors is also to the left of 0 and the other is to the right, since a negative times a positive equals a negative.

Q6 Many possible answers. In all cases, both a and b are between 0 and 1. When students explore this question, they may find it beneficial to show the number line controls and change the unit distance to 50 pixels.

DYNAMIC MULTIPLICATION

Q7 Both a and $a \cdot b$ remain fixed at 0. This makes sense because any number times 0 equals 0. (This is the multiplication property of zero.)

Q8 b and $a \cdot b$ stay right on top of each other. This makes sense because any number times 1 equals itself. (This is the multiplication property of one: the number 1 is the *multiplicative identity*.)

Q9 b and $a \cdot b$ move in opposite directions and at the same speed. They can be thought of as reflections of each other across 0. This makes sense because any number times -1 equals its opposite—the number just as "big" but on the other side of 0.

Q10 b moves faster than $a \cdot b$—twice as fast, to be exact. This is because a number multiplied by a number between 0 and 1 gives a result smaller (closer to 0) than the original number.

Q11 The two possible answers are 2 and -2.

WHY IS A NEGATIVE TIMES A NEGATIVE A POSITIVE?

Q12 When b is positive, dragging a to the left also moves $a \cdot b$ to the left. As a approaches 0, $a \cdot b$ approaches 0 too, and when $a = 0$, $a \cdot b = 0$. It makes sense that as you keep dragging a to the left, $a \cdot b$ continues its previous behavior, moving to the left into negative territory.

Q13 With b to the left of 0, a and $a \cdot b$ move in opposite directions. As you drag a to the left, $a \cdot b$ moves to the right. As a approaches 0, $a \cdot b$ approaches 0 too, but from the other side, and when $a = 0$, $a \cdot b = 0$. It makes sense that as you keep dragging a to the left, $a \cdot b$ continues its previous behavior, moving to the right into positive territory.

EXPLORE MORE

Encourage students to explore one or both of the geometric constructions. Students will find Object Properties (and the Parent/Child pop-up menus) useful in understanding the existing constructions.

WHOLE-CLASS PRESENTATION

This whole-class presentation works well if you have a student operate the computer while you provide direction, ask questions, and lead the discussion.

Start the presentation by opening the sketch **Multiplying Machine Present.gsp** and asking students how the markers show the original multiplication problem. Ask them to predict what will happen if you drag

ACTIVITY NOTES

marker *a* to the right, and what will happen if you drag it to the left. Drag the marker so they can test their predictions.

Continue by posing questions Q1–Q6 from the student activity sheet. (The presentation sketch has pages corresponding to each of the questions, although you may not need to use them.)

Pose questions Q7–Q11, paying particular attention to the machine's behavior when *a* has been dragged to values of 0, 1, and −1.

Finish the class discussion by analyzing what happens when you start with positive values for both *a* and *b,* and then slowly drag each in turn to the negative side of the number line. Ask students to explain in their own words how the machine's behavior makes sense when multiplying a positive times a negative (Q12), and why it makes sense when multiplying a negative times a negative (Q13). Elicit responses from a number of students, dragging the markers to illustrate their answers, so that various students get a chance to explain the machine's behavior in terms that make sense to them.

Exploring Algebra 1 with The Geometer's Sketchpad
© 2012 Key Curriculum Press

Raz's Magic Multiplying Machine

"Step right up, folks! Have I got a machine for you. You've seen number lines, right? I don't mind telling you, all those other number lines lack pizzazz . . . or, should I say, Raz-matazz?

"Say hello to my latest state-of-the-art number line. For starters, give it any two numbers and it multiplies them. Nifty, eh?

"But there's more. The eye-popping things this number line does will change the way you think about multiplication. Come give it a whirl!"

STATIC MULTIPLICATION

You can use the right and left arrow keys on your keyboard to move a selected marker one pixel at a time.

1. Open **Multiplying Machine.gsp.** Drag markers a and b to see how they behave and how they affect marker $a \cdot b$.

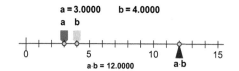

2. Drag markers a and b to represent the problem $3 \cdot 4 = 12$.

The product marker, $a \cdot b$, is to the right of the factors, a and b. This makes sense, since the product of two numbers greater than 1 is always bigger than either factor.

You can change the scale of the number line by showing the number line controls and either pressing the *Unit Distance* buttons or dragging point *scale*.

Q1 Drag a to 0.5 and b to 6. What does the machine show for the product $0.5 \cdot 6$?

Q2 List four pairs of locations for a and b such that $a \cdot b = 6$.

Q3 Find a location for a and b in which all three markers lie directly on top of each other. Is there more than one location that works?

Q4 If b and $a \cdot b$ are the same distance away from 0, but on opposite sides, where must a be?

Q5 Can a, b, and $a \cdot b$ all lie to the left of 0? Explain.

Q6 Find locations for a and b to the right of 0 such that $a \cdot b$ is smaller than both a and b. Describe all locations for a and b for which this works.

DYNAMIC MULTIPLICATION

When you multiply two numbers on a calculator, the only thing you see is the answer. Raz's machine gives the answer, but as you drag a or b toward its intended value, you get to observe the product, $a \cdot b$, as it moves simultaneously. This is fun to watch, and it can deepen your understanding of multiplication.

3. Drag a to 0. Then slowly drag b back and forth along the number line.

Q7 What happens to the product? Why does this make sense?

Q8 Drag a to 1. Then drag b back and forth along the number line. Describe the movement of $a \cdot b$ in relation to b. Why does this behavior make sense?

Q9 Drag a to -1. Then drag b back and forth along the number line. Describe the movement of $a \cdot b$ in relation to b. Why does this behavior make sense?

Q10 Drag a to 0.5. Then drag b back and forth along the number line. Which moves faster, b or $a \cdot b$? Explain why.

Q11 Find a location for point a such that the distance from $a \cdot b$ to 0 is always twice the distance from b to 0. Are there other answers?

WHY IS A NEGATIVE TIMES A NEGATIVE A POSITIVE?

Have you ever wondered why a negative number times a negative number is a positive number? Raz's machine provides a nice way to visualize the reason.

Remember you can use the arrow keys to drag a slowly.

4. Move both a and b so that they're near the right edge of the sketch window. Now drag a slowly to the left, and watch $a \cdot b$ glide across the screen.

When $a \cdot b$ reaches the left edge of the sketch, move a in the opposite direction so that $a \cdot b$ glides back to the right. Drag a back and forth while observing $a \cdot b$.

This type of explanation is sometimes called "reasoning by continuity."

Q12 Based on what you've observed, explain why it makes sense that a positive number times a negative number equals a negative number.

Q13 Move b to the left of 0. Once again, drag a back and forth, observing the behavior of the product, $a \cdot b$. Based on the behavior you observe, explain why it makes sense that a negative number times a negative number equals a positive number.

EXPLORE MORE

Go to the Construction pages and explore the first two designs Raz manufactured, using geometric technology. Try to duplicate his constructions. Then try to design and construct your own machine.

Multiple Models of Multiplication

This activity has two main purposes: to provide students with multiple models of multiplication and to give a variety of justifications for the rules for multiplying negatives.

By using multiple models of multiplication, students consider different ways of conceiving of this key operation and gain perspective on its meaning and uses. For this reason, don't allow students to do the problems in their heads without modeling them in Sketchpad—that would defeat the purpose.

As they work, students see how each model provides justification for the rules for multiplying negatives. A mental image provides a more solid foundation than a verbal rule for the idea that the product of two negative numbers is positive.

Keep in mind that these four models of multiplication aren't the only ones. The Raz's Magic Multiplying Machine activity provides yet another model that challenges students to broaden their thinking about multiplication and gives compelling reasons for the rules for multiplying negatives. These two activities work especially well together.

MULTIPLICATION AS JUMPING

Q1 Jumps that end up at 6 include $3 \cdot 2, 2 \cdot 3, 6 \cdot 1, 1 \cdot 6, -3 \cdot (-2)$, $-2 \cdot (-3), -6 \cdot (-1)$, and $-1 \cdot (-6)$.

Q2 When the number of jumps is negative and each jump is positive, the rabbit faces right and jumps backward, moving to the left. When the number of jumps is positive but each jump is negative, the rabbit faces left and jumps forward, again moving to the left.

Q3 When both the number of jumps and the size of each jump are negative, the rabbit faces left and jumps backward, moving to the right. He faces left because the size of the jumps is negative, and he jumps backward because he's taking a negative number of jumps. By facing left and jumping backward, the rabbit moves in the positive direction along the number line.

MULTIPLICATION AS GROUPING

Q4 a. $4 \cdot 2 = 8$ b. $3 \cdot (-3) = -9$

c. $1 \cdot (-8) = -8$ d. $8 \cdot (-1) = -8$

e. $-2 \cdot 3 = -6$ f. $-1 \cdot 5 = -5$

g. $-2 \cdot (-3) = 6$ h. $-8 \cdot (-1) = 8$

Q5 a. $3 \cdot (-4) = -12$ b. $4 \cdot (-3) = -12$

These two results are similar in that they both give the same negative answer, -12. In both cases, one number is positive and one is negative. The biggest difference is that the reason for changing direction, from positive to negative, is completely different in the two cases.

Q6 "Take away three groups of 4" ($-3 \cdot 4 = -12$) and "take away four groups of 3" ($-4 \cdot 3 = -12$).

MULTIPLICATION AS AREA

Q7 When the width becomes negative, the rectangle flips over horizontally, the squares change color, and the area becomes negative. Some students may make a valuable logical connection between the flipping of the rectangle and the area becoming negative.

Q8 $-1 \cdot 6 = -6$ $-2 \cdot 3 = -6$

$-6 \cdot 1 = -6$ $1 \cdot (-6) = -6$

$2 \cdot (-3) = -6$ $3 \cdot (-2) = -6$

$6 \cdot (-1) = -6$

Q9 $1 \cdot 4 = 4$ $2 \cdot 2 = 4$

$4 \cdot 1 = 4$ $-1 \cdot (-4) = 4$

$-2 \cdot (-2) = 4$ $-4 \cdot (-1) = 4$

Q10 Every square number can be modeled with a square in the area multiplication model. For example, 4 can be modeled by $2 \cdot 2$ or $-2 \cdot (-2)$, both of which are squares.

(A number such as -4 can also be modeled with squares, $2 \cdot (-2)$ or $-2 \cdot 2$. However, in these squares, the base and height are not equal. This can be interpreted as a weakness of this model, or it might represent an opportunity for a sneak preview of imaginary numbers.)

MULTIPLICATION AS SCALING

Q11 a. The mapping segments point straight down, parallel to each other. Every number maps to itself. For example, $2 \cdot 1 = 2$, $-3 \cdot 1 = -3$, $0 \cdot 1 = 0$, etc.

b. The mapping segments point inward toward the bottom. Every number maps to a number whose absolute value is less than its own absolute value (or equal to, in the case of 0), but whose sign is the same. For a scale factor of 0.5, for example, $2 \cdot 0.5 = 1$, $-3 \cdot 0.5 = -1.5, 0 \cdot 0.5 = 0$, etc.

c. The mapping segments all point to zero, so every number maps to zero. For example, $2 \cdot 0 = 0, -3 \cdot 0 = 0, 0 \cdot 0 = 0$, etc.

d. The mapping segments cross between the two number lines. Every number maps to a number with the opposite sign (except for 0, which points to itself). For a scale factor of -2, for example, $2 \cdot (-2) = -4, -3 \cdot (-2) = 6, 0 \cdot 0 = 0$, etc.

Q12 a. $a = 0.25$ b. $a = -2$

      c. $a = -1$ d. $a = -0.1$

Q13 a. $b = 4; a = 1/4$ b. $b = -1/2; a = -2/1$

      c. $b = -1/1; a = -1/1$ d. $b = -10; a = -1/10$

In each pair, the numbers are reciprocals of each other. For example, in part a, $b = 4/1$ and $a = 1/4$.

SUMMING UP

Q14 There are many possible answers. We feel that Jumping and Grouping are particularly effective as an introduction to multiplication. They correspond with most people's basic conception of multiplication and so are a good place to start. Area is particularly effective at showing the "dimensionality" of multiplication—how multiplying two one-dimensional objects produces a two-dimensional object. Scaling is good for showing how multiplication affects an entire set of objects, including non-integers. It also serves as a great introduction to dynagraphs.

Q15 Jumping, Grouping (especially when using the terms "put together" and "take away"), and Scaling are effective at demonstrating the rules of multiplication for negatives. Area is less effective for this, in our view, because there is no compelling reason why the rectangles in the first and third quadrants are blue and those in the second and fourth quadrants are red.

EXPLORE MORE

Q16 Students should model pairs of equations, such as $2 \cdot (-5) = -10$ and $-5 \cdot 2 = -10$. Area may be especially useful for demonstrating commutativity because it's so easy to see that the two rectangles have the same area and sign.

WHOLE-CLASS PRESENTATION

The whole-class presentation of this activity substantially follows the steps of the student activity sheet. Refer to the Presenter Notes for tips to follow and adjustments to make so that the presentation can be as useful to students as possible.

Multiple Models of Multiplication

You can present any of the models in this activity independently, though it's valuable to present at least two models in succession. Students will get the greatest benefit from this activity when they compare the behaviors of several different models.

Follows the steps in the student activity sheet, with the adjustments described below.

MULTIPLICATION AS JUMPING

Be sure to elicit answers from a number of students.

When the rabbit first jumps, ask students how the rabbit's motion illustrates the multiplication problem shown before adjusting the numbers (leaving both positive) and doing another example.

Before changing the number of jumps to be negative, ask students to predict what the rabbit will do.

Be sure to make the jumps value positive again before you make the units value negative.

Similarly, before making the size of each jump negative, ask students to predict what the rabbit will do. And ask again for predictions before making both numbers negative at the same time.

MULTIPLICATION AS GROUPING

In the grouping part of the presentation, ask students to make up problems using particular combinations of negative and positive ("put together" and "take away") rather than using the specific ones from Q4. Be sure to show how a "put together" problem and a "take away" problem can give the same result.

MULTIPLICATION AS AREA

When presenting multiplication as area, you may want to emphasize that the color of the rectangle indicates the sign of the result, without too much emphasis on the idea of "negative area." Don't let a discussion of negative area become a distraction.

MULTIPLICATION AS SCALING

Unlike the other models, the numbers don't need to be integers, so this page shows a continuous model of multiplication.

CONCLUSION

Finish the class discussion by asking students to compare the various models, particularly with regard to how they show the product of two negative numbers.

Multiple Models of Multiplication

What does *multiplication* mean? This question has many answers, because there are many ways of thinking about multiplication. In this activity you'll compare four such ways—multiplication as jumping, as grouping, as area, and as scaling.

MULTIPLICATION AS JUMPING

You can think of multiplication as jumping: Three jumps of two units each could be described by the multiplication problem 3 · 2. In this model, you will experiment with setting the number of jumps and the size of each jump.

1. Open **Multiplication Models.gsp.** Press the *Jump!* button to animate three jumps of two units each.

2. Press the *Reset* button, drag the circles to represent two jumps of five units each, and press the *Jump!* button again.

Q1 How many ways can you do jumps that end up at 6? Drag the green circles to try each way, and write down all the ways you found.

Q2 Change the number of jumps so that it's negative. What happens during the jumping? Make the number of jumps positive again, and make the size of each jump negative. What happens?

Q3 What happens if the number of jumps and the size of each jump are both negative? How can you explain this logically?

MULTIPLICATION AS GROUPING

You can also think of multiplication as grouping: 3 · 2 means three groups of two things each. In this model, you will group rectangles along a number line.

3. Go to the Grouping page. The objects in the sketch model the sentence "Put together three groups of two." The equation is 3 · 2 = 6.

Q4 Drag the circles to model each sentence below. On your paper, draw the bottom shape (the one on the number line) and write its equation.

    a. Put together four groups of 2.   b. Put together three groups of −3.

    c. Put together one group of −8.   d. Put together eight groups of −1.

How should you drag the top circle to represent "take away"?

e. Take away two groups of 3.   f. Take away one group of 5.

g. Take away two groups of -3.   h. Take away eight groups of -1.

Q5 Model the following sentences and write their equations. How are they similar and how are they different?

   a. Put together three groups of -4.   b. Put together four groups of -3.

Q6 Using 4's and 3's, write and model two "take away" sentences whose product is the same as the product in Q5.

MULTIPLICATION AS AREA

Another way to think about multiplication is in connection with the area of rectangles.

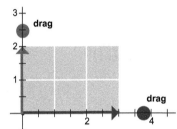

4. Go to the Area page. In this model, the height and width can be either positive or negative. When you start, both are positive.

Q7 Drag the width to model $-3 \cdot 2 = -6$. What happens to the rectangle when the width becomes negative? What does this change indicate about the area?

Q8 Model seven different problems in which the area equals –6. Write the problems on your paper.

Q9 Model and write down as many problems as you can in which the area equals 4.

Q10 The numbers $1, 4, 9, 16, \ldots$ are called "squares." Explain why this makes sense given the area model of multiplication.

MULTIPLICATION AS SCALING

Whether you're drawing a scale model of your room or scaling a recipe to serve more people, you're using multiplication.

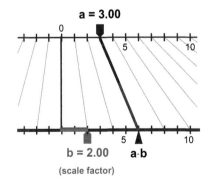

5. Go to the Scaling page. The *scale factor* (*b*) is 2, so every number on the top axis *maps* (corresponds) to a number twice as big on the bottom axis. Point *a* is at 3 and maps to 6, so the equation for this problem is $3 \cdot 2 = 6$.

6. Drag *a* to model $-3 \cdot 2 = -6$. Then drag *b* to model $-3 \cdot (-2) = 6$.

Q11 Describe what the gray mapping segments look like when:

a. *b* equals 1.

b. *b* is between 0 and 1.

c. *b* equals zero.

d. *b* is negative.

Q12 For each problem below, set the scale factor *b* as listed, and then drag *a* so that $a \cdot b = 1$. (For example, if b were 0.5, you would make $a = 2$ because $0.5 \cdot 2 = 1$.)

a. $b = 4$; $a \cdot b = 1$; $a = ?$

b. $b = -0.5$; $a \cdot b = 1$; $a = ?$

c. $b = -1$; $a \cdot b = 1$; $a = ?$

d. $b = -10$; $a \cdot b = 1$; $a = ?$

Q13 Rewrite the answers to Q12 using fractions instead of decimals. What do you notice?

SUMMING UP

Q14 List one strength of each of the four models, perhaps something that each shows about multiplication better than the others.

Q15 Which of the four models do you think is most effective at showing why the product of two negatives is a positive? Defend your choice.

EXPLORE MORE

To copy from Sketchpad into a word processor, select the objects you want to copy, and resize the window to the desired size of your picture. Then choose **Edit | Copy,** and paste the result into your word processor.

Q16 The commutative property of multiplication says that it doesn't matter whether you multiply $3 \cdot 2$ or $2 \cdot 3$; you get the same answer using either order. Set up four pairs of multiplication problems, one for each model, to show this property. Copy the Sketchpad image for each of the eight problems and paste them into a word processor document. Which model do you think is most effective at showing why multiplication is commutative?

Mystery Machines

CHALLENGES

Q1 To find 0, drag *a* (or *b*) until it and *a* · *b* are right on top of each other. This spot is 0. This method works because any number times 0 equals 0. Thus, when *a* is 0, *a* · *b* will also be 0. Once you've positioned *a*, you can check your work by dragging *b*. If *a* is at zero, then *a* and *a* · *b* will remain on top of each other as *b* is dragged.

To find 1, drag *a* until *b* and *a* · *b* are right on top of each other (or drag *b* until *a* and *a* · *b* are right on top of each other). *a* is at 1. This method works because any number times 1 equals itself. Thus, when *a* is 1, *b* and *a* · *b* will equal each other. Again, check your work by dragging. If *a* is 1, then *b* and *a* · *b* will remain on top of each other as *b* is dragged.

Q2 Move *a* and *b* so that they're both at 1/2. (1/2)(1/2) = 1/4, so *a* · *b* is at 1/4. Mark this spot using one of the gold arrows. Now move either *a* or *b* to 1/4 (where you just marked). *a* · *b* will now be at 1/8 because (1/2)(1/4) = 1/8.

Q3 To find 0, drag *a* until *b* and *a* + *b* are right on top of each other (or drag *b* until *a* and *a* + *b* are right on top of each other). This works because any number plus 0 equals itself. Thus, when *a* is 0, *b* and *a* + *b* will equal each other. (Note that the method for finding 0 on an "adding machine" is the same as that for finding 1 on a "multiplying machine." This is because 1 is the *identity element* for multiplication and 0 is the *identity element* for addition.)

It's impossible to find 1, because there's no way to find the scale using addition.

Q4 Mystery combo 1: $c = a + 2b$

Mystery combo 2: $c = 4a - b$

Mystery combo 3: $c = ab - 1$

EXPLORE MORE

Q5 This challenge involves creating an algebraic construction. This is somewhat easier to build than the geometric construction, and the algebraic approach allows the machine to perform much more complex computations.

WHOLE-CLASS PRESENTATION

Use the Challenges in this activity to encourage a vibrant class discussion. As the class looks at each challenge, have students take turns operating the computer while other students make suggestions about how to investigate each machine. The challenge posed by Machine 3 is particularly interesting because one of the questions has no solution: It's not possible to find the position of 1. (Sometimes a result of impossibility is more enlightening than a more straightforward challenge with a well-defined answer.)

When students work with the Combo pages, be sure to have them explain and defend their strategies. Emphasize how important it is that they be able to explain their thought processes, and point out that they don't fully understand a problem until they can explain its solution to someone else.

The Explore More section is not particularly suited for use in a whole-class presentation.

Mystery Machines

Raz is an inventor who got started building multiplying machines and then moved on to build addition, subtraction, and division machines. Now he has decided to move from the arithmetic market into the more lucrative game market by building mystery machines. In this activity, you will explore some of his new machines.

CHALLENGES

1. Open **Mystery Machines.gsp.**

The first machine is an ordinary multiplying machine, but with the numbers left off. Follow the directions in the sketch to find the locations of 0 and 1. Use the *New problem* button to try the challenge several times.

Q1 Describe the strategies you used to find the locations of 0 and 1.

2. On the second page (Machine 2), the challenge is to find and mark the location of the number 1/8. Repeat this challenge several times.

Q2 Describe your strategy clearly.

3. The third page (Machine 3) takes two numbers, a and b, and computes their sum, $a + b$.

Q3 Can you find 0? If so, describe your strategy; if not, explain why not. Can you find 1? If so, describe your strategy; if not, explain why not.

> Look for clues to help you determine what rule is being used to calculate *c*.

4. Each Combo page contains a new machine that takes two numbers, a and b, and uses them to compute a third number, c. The machine uses a formula like $c = a + b + 2$, or $c = b - a$, or $c = 2a + b$. Investigate each machine by dragging a and b and observing the effects on c.

Q4 For each of the three pages, tell what formula is being used, and describe your strategy for discovering the formula.

EXPLORE MORE

> Raz's Combo machines can give you some ideas for inventing your own machine.

Q5 The Build It Yourself page gives directions for building your own math machine. Follow the directions to build a division machine. Then change the calculation to build a machine of your own invention.

Dividing Real Numbers

MULTIPLICATION MACHINE

Q1 Marker *a* determines how many rectangles appear in the blue bar. Marker *b* determines the width of each blue rectangle.

Q2 When *a* shows exactly one yellow square, markers *b* and *c* move in unison.

Q3 You can calculate the length of the row of blue tiles by multiplying the number of tiles by the length of each tile. As an equation, $c = a \cdot b$.

DIVISION MACHINE

8. This is the simplest step of the activity, but in many ways the most important. Encourage students to observe closely and to think about what they observe.

Q4 When students press the button, the control for *b* goes below the line and the control for *c* goes above it, and students find that they can now control *c* rather than *b*. This is what makes it a division machine now: It calculates $b = c \div a$.

Geometrically, students are now controlling the size of the blue rectangles by dragging the end of the bar (representing the product) rather than by dragging the end of the first rectangle (the multiplier).

Q5 You would divide *c* by *a*. $b = c \div a$

Q6 $b = 5 \ (15 \div 3 = 5)$

$b = 3 \ (7.2 \div 2.4 = 3)$

Q7 Emphasize to students the importance of these four questions in developing and demonstrating their skills of translation and interpretation. Students must translate the questions, which are in mathematical terms, into the terms of the model and the behavior of the markers; must manipulate the model appropriately; and then must interpret the behavior they observe, expressing their conclusions mathematically.

   a. When you drag *a* to exactly one, markers *b* and *c* are at the same value, no matter where you drag *c*. Dividing by one does not change the number.

   b. No. When you drag *a* to be negative (for example, −2) and also drag *c* to be negative (for example, −6), the result is positive. As an equation, $-6 \div (-2) = 3$. Dividing by a negative number always

Exploring Algebra 1 with The Geometer's Sketchpad
© 2012 Key Curriculum Press

changes the sign, so dividing a negative by another negative gives a positive answer.

c. No. When you drag a between zero and one, the result is that b is larger than c. This means that dividing a positive number by a number between zero and one results in a larger number.

d. When you drag a toward zero while b is positive, the marker for b moves off the screen, showing a larger and larger result. When you get a to exactly zero, the machine breaks and the entire blue bar disappears. If you try the same thing while c is already zero, b also stays at zero as a gets close to zero, but when a is exactly zero, the machine breaks and b disappears.

Algebraically, if $c \div 0 = b$, then $c = 0 \cdot b$. If c is not zero, then there is no number b that will satisfy the second equation, because $0 \cdot b = 0$. On the other hand, if $c = 0$, then any real number will do for b, so there is no unique answer.

WHOLE-CLASS PRESENTATION

To present this activity to the whole class, start on page 2 of the sketch **Division Machine.gsp** and drag markers a and b to see how the machine performs multiplication. Ask students to explain the role of each variable in determining how the machine works. It's helpful to relate the behavior of this machine to the grouping model in the Multiple Models of Multiplication activity.

Once students are satisfied with the multiplying machine, press the *Multiply/Divide Toggle* button and observe the changes in the sketch. Switch this toggle several times to make sure students have seen the changes clearly, and then observe the behavior of the division machine's markers. As the class explores the parts of Q7, it may be helpful to switch back and forth several times between multiplication and division.

Dividing Real Numbers

When you first learned about division, your teacher probably began with a problem about distributing something fairly, such as "Divide 12 marbles among three children." Problems like this go only so far, because marbles and children come in positive integers. Division has to work for negative numbers and fractions.

MULTIPLICATION MACHINE

To understand division, you must first understand multiplication, so this activity starts with a multiplication machine.

1. Open **Division Machine.gsp.**

   Q1 Experiment by dragging markers *a* and *b*. How can you control the number of rectangles in the blue bar? How can you control their width?

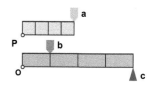

   Q2 How do *b* and *c* behave when *a* is set to show exactly one yellow square?

To do real multiplication and division, you should have numbers, and for that you need a number line. You can follow steps 2–6 to make your own number lines, or you can go to page 2 of the sketch and skip to Q3 on the next page.

To merge the points, use the **Arrow** tool to select both of them and then choose **Edit | Merge Points.**

2. Press the *Show Number Line* button. Attach the blue bar to the number line by merging point *O* on the blue bar and point *O* on the number line.

You'll need a separate number line for the yellow bar so that the bars don't overlap.

3. Construct a line through point *O*, perpendicular to the existing number line, by using the **Arrow** tool to select point *O* and the line and then choosing **Construct | Perpendicular Line.**

4. Use the **Number Line** custom tool to create the second number line. Press and hold the **Custom** tool icon, and choose **Number Line** from the menu that appears. Use the tool by clicking twice: first on the *Unit Distance* measurement and then on the perpendicular at the position where you want the new number line to appear.

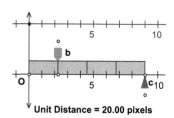

Unit Distance = 20.00 pixels

5. Choose the **Arrow** tool, and attach the yellow bar to the new number line by merging point *P* with the zero point of the number line.

Exploring Algebra 1 with The Geometer's Sketchpad
© 2012 Key Curriculum Press

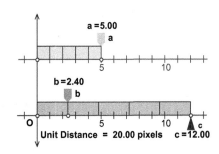

To hide points, use the **Arrow** tool to select them, and then choose **Display | Hide Points.**

6. To measure the position of each marker on the number line, use the **Measure Value** custom tool. Click each of the three points that appear above or below markers *a, b,* and *c.* Hide the three points.

Q3 If you know the length of one blue tile and the number of tiles, you should be able to calculate the length of the row of blue tiles. How would you do this calculation? Write your answer as an equation using *a, b,* and *c.*

Choose **Number | Calculate.** Click the measurements to enter them in the calculation.

7. Use the measurements of *a* and *b* to calculate $a \cdot b$. Drag *a* and *b* to make sure that your calculation is correct.

DIVISION MACHINE

8. Press the *Multiply/Divide Toggle* button.

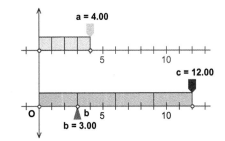

Q4 What changed when you pressed the button? Which markers can you control now? How is the behavior of the machine different?

If you can't get the exact numbers you want, click the *Round* button for the value you're moving, and then try again.

Q5 By construction, $c = a \cdot b$. If you know the values of *a* and *c*, how would you use those numbers to calculate *b*?

Q6 Use the division machine to calculate $15 \div 3$. (To do this, drag *c* to 15 and drag *a* to 3.) What is *b*? Then use the same method to calculate $7.2 \div 2.4$.

Q7 Use your division machine to answer the following questions. For each question, tell how you dragged markers *a* and *c* to investigate, describe your observations, and explain your answers.

   a. What is the result when you divide a number by one?

   b. If you divide by a negative number, is the answer always negative?

   c. If you divide a positive number, is the answer always a smaller number?

   d. What happens when you try to divide by zero?

The Commutative Property

This activity and the one on the associative property provide an enjoyable way for students to investigate a bit of algebra theory.

ADDITION AND SUBTRACTION

Q1 The results of the two addition problems are always the same. Therefore addition is commutative.

Q2 The results of the two subtraction problems are the same only when the two operands are equal; under any other circumstances, the results are different. Subtraction is *not* commutative. The results $a - b$ and $b - a$ are opposite.

MULTIPLICATION AND DIVISION

Q3 The length of the blue bar represents a multiplication problem in which the length of an individual rectangle is multiplied by the number of rectangles to get the total bar length. Following is the table filled in. (Answers will vary in the last two columns.)

	Marker	Value 1	Value 2	Value 3
Number of upper blue rectangles	a	2	2	−2
Length of each upper blue rectangle	b	6	5	3
Total length of upper blue bar	$a \cdot b$	12	10	−6
Number of lower blue rectangles	b	6	5	3
Length of each lower blue rectangle	a	2	2	−2
Total length of lower blue bar	$b \cdot a$	12	10	−6

Q4 No matter how the operands are changed, the two multiplication problems give the same answer. Multiplication is commutative.

Q5 The length of the blue bar represents a division problem in which the entire length (the dividend) is divided into a number of equal pieces. The number of pieces is the divisor, and the length of each individual piece is the quotient. Following is the table filled in. (Answers will vary in the last two columns.)

	Marker	Value 1	Value 2	Value 3
Total length of upper blue bar	a	6	−8	5
Number of upper blue rectangles	b	4	4	5
Length of each upper blue rectangle	a/b	1.5	−2	1
Total length of lower blue bar	b	4	4	5
Number of lower blue rectangles	a	6	−8	5
Length of each lower blue rectangle	b/a	0.67	−0.5	1

Q6 The two red markers (the quotients of the two division problems) represent equal quotients only when the absolute values of the dividend and divisor are equal. For instance, if a and b are both 3, then $a/b = b/a = 1$, or if $a = 4$ and $b = -4$, then $a/b = b/a = -1$. Because the results are not always equal, division is not commutative.

ALL TOGETHER

Q7 The blue bars represent the result of adding the two operands a and b. The first blue bar shows the result for $a + b$, and the second one shows the result for $b + a$.

Q8 As the operands are dragged, the two bars for addition and multiplication are always equal in length, indicating that these two operations are commutative. The green bars for the two subtraction results move in opposite directions, as do the magenta bars for division. Thus subtraction and division are not commutative.

WHOLE-CLASS PRESENTATION

Use the pages of **Commutative Property.gsp** to investigate whether each of the arithmetic operations is commutative. The Multiplication and Division pages are easier to understand if students are familiar with the grouping models presented in Multiple Models of Multiplication and in Dividing Real Numbers. Get a number of students to formulate their conclusions in their own words to make sure that all of them understand what commutativity means and why some operations are commutative and others are not.

The Commutative Property

At a dinner table, you might ask for the salt and pepper, but you could ask for the pepper and salt and expect the same outcome. But sometimes the order of things is important. You can put on shoes and socks, but you'd better start with the socks.

An operation that is commutative is said to have the *commutative property.*

In each of the four elementary arithmetic operations, there are three numbers: the two you start with (called *operands*) and the result. Does it matter which operand comes first? If changing the order of the operands doesn't change the result, the operation is *commutative.*

ADDITION AND SUBTRACTION

1. Open **Commutative Property.gsp.** The top blue and green arrows are the operands. Drag the points at their tips to change their values.

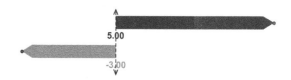

The top number line shows the addition in the order *blue + green.* The bottom number line shows the addition in the opposite order: *green + blue.*

For addition to have the commutative property, the two results must always be the same.

Q1 Are the results of the two addition problems (*blue + green* and *green + blue*) ever the same? Are they always the same? Drag the points to try many combinations of values. Does addition have the commutative property?

2. Go to the Subtraction page. Notice that in actually doing the subtraction problems, the bottom arrow has been flipped in order to subtract the second operand from the first.

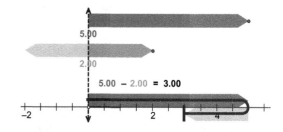

Q2 Experiment by changing the values. When is *blue − green* equal to *green − blue*? Are they always equal? Is subtraction commutative?

MULTIPLICATION AND DIVISION

You may have seen this model in a previous activity.

3. Go to the Multiplication page.

Q3 How does each blue bar correspond to a multiplication problem? Make a copy of the table below and finish filling in the first two columns for the blue bars.

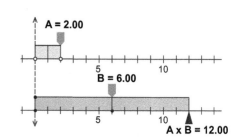

The entries in the first row indicate that marker *a* corresponds to the number of rectangles in the upper blue bar, and that its initial value is 2.

	Marker	Value 1	Value 2	Value 3
Number of upper blue rectangles	*a*	2		
Length of each upper blue rectangle				
Total length of upper blue bar				
Number of lower blue rectangles	*b*	6		
Length of each lower blue rectangle				
Total length of lower blue bar				

Q4 Create two more problems by moving markers *a* and *b*. Use your results to fill in the remaining columns of the table. When do the two red markers (showing the lengths of the two blue bars) represent equal products? Are they always equal? Is multiplication commutative?

4. Go to the Division page.

Q5 How does each blue bar correspond to a division problem? Copy this table and finish filling in the first two columns.

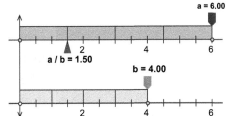

	Marker	Value 1	Value 2	Value 3
Total length of upper blue bar	*a*	6		
Number of upper blue rectangles				
Length of each upper blue rectangle				
Total length of lower blue bar	*b*	4		
Number of lower blue rectangles				
Length of each lower blue rectangle				

Q6 Create two more problems by moving markers *a* and *b*. Use your results to fill in the remaining columns of the table. When do the two red markers (showing the lengths of the individual rectangles in the upper and lower blue bars) represent equal quotients? Are they always equal? Is division commutative?

ALL TOGETHER

5. Go to the Summary page. The red bars represent the operands, *a* and *b*. Experiment with different values by dragging the points at their tips.

Q7 What do the blue bars represent? Why are there two of them?

Be sure to drag each operand separately. Don't drag them both at the same time.

Q8 Drag each operand in turn to change its value. Which of the four operations are commutative? How can you tell this from the behavior of the bars?

The Associative Property

This activity provides an enjoyable way for students to investigate a bit of algebra theory.

ADDITION

2. Students first use a custom tool in step 2. If they are not experienced with custom tools, there may be a tendency to leave the tool activated and inadvertently click on the screen, creating unwanted objects. Tell them to choose the **Selection Arrow** as soon as they finish step 3.

Q1 The values $(a + b) + c$ and $a + (b + c)$ are equal for any a, b, and c.

Q2 Addition does have the associative property. Students' explanations will vary. One explanation that helps them remember the meaning of this property is the following: "When you add three numbers, it doesn't matter whether the second number associates with the first number or whether it associates with the third number. Whichever number it adds itself to, the final result will be the same."

OTHER OPERATIONS

Q3 Addition and multiplication have the associative property. Subtraction and division do not.

EXPLORE MORE

Q4 If students have trouble with this question, consider giving them a hint. In mathematics, special things tend to happen around the numbers zero and one.

For subtraction, even without associativity, the equation is satisfied when $c = 0$. For division, the equation is satisfied for any one of these conditions: $a = 0$, $c = 1$, or $c = -1$.

WHOLE-CLASS PRESENTATION

Use the pages of **Associative Property.gsp** to investigate whether each of the arithmetic operations is associative. The presentation requires the use of custom tools. If you're new to using custom tools, practice the steps described in the activity sheet before presenting to the class.

As you present this activity, emphasize that you follow a different order when calculating the upper answer and the lower answer, and then check whether the two answers are the same regardless of order. In the discussion it's particularly helpful to have a number of students describe in their own words what it is about the behavior of the markers that allows them to decide whether a particular operation is associative.

Imagine that you've bought three items at a store and that you're adding up the three prices (call them a, b, and c) to check the bill. Does it matter whether you first add a and b, and then add c to the result, or could you instead add b and c first, and then add the result to a? Does $(a + b) + c = a + (b + c)$?

Instead of adding, what if you were subtracting the three numbers? Does $(a - b) - c = a - (b - c)$? What if you were multiplying or dividing them? Would it matter in these cases whether you performed the first calculation on a and b, or whether you performed it on b and c?

To write the question so it applies to any of the four arithmetic operations, use $\otimes$ to stand for the operator. You want to know if you'll get the same result by doing $a \otimes b$ first as you will if you do $b \otimes c$ first. Using parentheses to show the order, you want to know if

$$(a \otimes b) \otimes c = a \otimes (b \otimes c)$$

Another way of phrasing the question is to ask whether the b should be *associated* with the a or the c. If you get the same answer either way, the operation is called *associative* and is said to have the *associative property*. In this activity you will investigate this question for addition, subtraction, multiplication, and division to determine which of these operations are associative and which are not.

ADDITION

Begin by checking whether addition has the associative property. You must decide whether the following equation is true for all real numbers a, b, and c:

$$(a + b) + c = a + (b + c)$$

1. Open **Associative Property.gsp.**

There are two number lines with markers. For now, work on the top line.

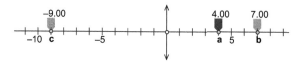

2. Press and hold the **Custom** tool icon to display the Custom Tools menu. Choose the **Add** tool. Click in order on points a and b. (You can use the **Text** tool to center the label of $(a+b)$ above its marker.)

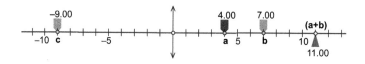

Exploring Algebra 1 with The Geometer's Sketchpad
© 2012 Key Curriculum Press

3. With the **Add** tool still active, click in order on point $(a + b)$ and point c.

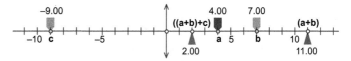

Choosing the **Selection Arrow** tool deactivates the **Add** tool.

4. Use the **Arrow** tool to drag points a, b, and c and confirm that the new markers, representing $(a + b)$ and $(a + b) + c$, show the correct values.

5. On the lower number line, use the **Add** tool to construct $(b + c)$. Then construct $(a + (b + c))$.

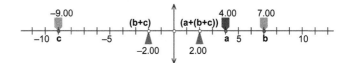

You can drag the two number lines closer together for a better comparison. Change the scale by dragging the tick-mark numbers on L_1.

Q1 Compare the final values created on the two number lines. Drag a, b, and c to try different values. Do $(a + b) + c$ and $a + (b + c)$ always have the same value?

Q2 Does addition have the associative property? Explain in your own words what this means.

OTHER OPERATIONS

Finish the investigation on your own, using the "Subtraction", "Multiplication", and "Division" pages. There is a custom tool corresponding to each of the operations. In your investigation, decide which of these equations are true:

$$(a - b) - c = a - (b - c)$$

$$(a \cdot b) \cdot c = a \cdot (b \cdot c)$$

$$(a \div b) \div c = a \div (b \div c)$$

Q3 Of the four elementary arithmetic operators, which have the associative property and which do not? Illustrate each conclusion with an example.

EXPLORE MORE

In each section of this investigation, you had to try many different values for the given points. Even if an operation does not have the associative property, there may be some special combination of values for a, b, and c for which the equation is true.

Q4 Go back to the operations that are not associative. Using your construction, find a case in which the equation is true even though there is no associative property.

Identity Elements and Inverses

This activity provides an enjoyable way for students to investigate a bit of algebra theory.

This activity gives detailed directions for investigating the identity element and inverses for addition, and then has students perform their own investigations for the other operations.

IDENTITY ELEMENT FOR ADDITION

Q1 The $e + x$ marker appears at the same spot as the $x + e$ marker, no matter where you drag x or e. This occurs because addition is commutative.

Q2 Addition does have an identity element—the number 0.

INVERSES FOR ADDITION

Q3 The markers for $x + y$ and $y + x$ come out at the same place, because addition is commutative.

Q4 Recorded values of x and y will vary, but in each case, the numbers should have opposite signs and identical magnitudes.

IDENTITY ELEMENT AND INVERSES FOR OTHER OPERATIONS

Q5 If an operation doesn't have an identity element, there can be no inverses with respect to that operation, because the definition of an inverse depends on the value of the identity element.

Q6 Actual numbers will vary. Here's a typical table:

Operator	Identity Element	Examples of Inverses
Addition	0	$(3, -3), (-5, 5), (1.5, -1.5)$
Subtraction	none	none
Multiplication	1	$(2, 0.5), (-0.25, -4), (10, 0.1), (-3, -1/3)$
Division	none	none

Q7 The number zero has no inverse with respect to multiplication. You may want to ask students how their answer to this question will change if they limit themselves to the set of positive integers and zero, or to the full set of integers.

Exploring Algebra 1 with The Geometer's Sketchpad
© 2012 Key Curriculum Press

In the case of positive integers and zero, the number 0 is the only number that has an inverse with respect to addition, and the number 1 is the only number that has an inverse with respect to multiplication.

In the case of the full set of integers, every number has an inverse with respect to addition, but only the number 1 has an inverse with respect to multiplication.

WHOLE-CLASS PRESENTATION

Use the pages of **Identity Elements.gsp** to investigate whether each of the arithmetic operations has an identity element and inverses. The presentation requires the use of custom tools, and is best done after you and your students have performed at least one other activity using custom tools.

As you present this activity, have a different student operate the computer as you investigate each of the four operations. In the discussion of each of the four conclusions, it's particularly helpful to have several students describe in their own words why the behavior of the markers for a particular operation shows or does not show an identity element and inverses. Be sure to include Q7 in the discussion, so that students realize that some numbers may not have inverses even when there's an identity element.

Identity Elements and Inverses

This activity introduces the concepts of identity elements and inverses. You'll determine whether addition, subtraction, multiplication, and division have identity elements. For any that do have identity elements, you'll investigate whether particular values have inverses.

IDENTITY ELEMENT FOR ADDITION

An operation $\otimes$ has an identity element e if, for all possible values of x, $x \otimes e = e \otimes x = x$.

1. Open **Identity Elements.gsp.**

If an identity element e exists for addition, then $x + e = e + x = x$ for all possible values of x. On the upper number line, there are adjustable markers for x and e. You must create markers for $x + e$ and for $e + x$.

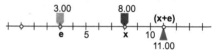

2. Press and hold the **Custom** tool icon to display the Custom Tools menu. Choose the **Add** tool. Click in order on points x and e.

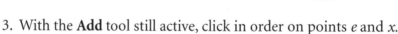

3. With the **Add** tool still active, click in order on points e and x.

Q1 Where does the $e + x$ marker appear relative to the $x + e$ marker? Does this relationship continue when you drag points x and e? What property of addition does this relationship represent?

You can use the **Text** tool to separate the labels for $(x+e)$ and $(e+x)$ so they don't overlap.

So far, you have determined that $x + e = e + x$. But there's one more requirement for e to be the identity element for addition: Both of these quantities must be equal to x.

4. Drag e until all three quantities (x, $x + e$, and $e + x$) are equal.

5. Drag x to determine whether this identity relationship is true for all values of x. Be sure to try negative values and values near zero.

Q2 If $x + e = e + x = x$ is true for all values of x, then e is the identity element for addition. Does addition have an identity element, and if so what is it?

Exploring Algebra 1 with The Geometer's Sketchpad
© 2012 Key Curriculum Press

INVERSES FOR ADDITION

If an operation has an identity element e, then particular numbers may also have inverses with respect to that operation. An inverse of x with respect to $\otimes$ is defined as a value y such that $x \otimes y = y \otimes x = e$.

Addition has an identity element. Find out whether numbers have inverses with respect to addition. For any particular value of x, you're looking for a value y such that $x + y = y + x = e$.

Be sure marker e remains at the position of the identity element for addition.

6. On the lower number line, use the **Add** tool to add $x + y$, and use it again to add $y + x$.

Q3 Where do $x + y$ and $y + x$ come out in relation to each other? Why?

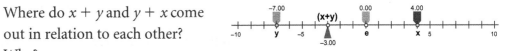

Q4 Use the **Arrow** tool to drag y so that $x + y = e$, and record the values of x and y. Then move x to a new position, drag y again so that $x + y = e$, and record these values. Repeat until you have four pairs of values for x and y. What can you conclude about inverses with respect to addition?

IDENTITY ELEMENT AND INVERSES FOR OTHER OPERATIONS

Q5 If a particular operation doesn't have an identity element, numbers cannot have inverses with respect to that operation. Why?

Q6 Repeat your investigation for other operations, using the remaining pages of **Identity Elements.gsp.** Record your results in a table like the one below. If an operation has no identity element, write "none" in the appropriate box.

Operator	Identity Element	Examples of Inverses
Addition		
Subtraction		
Multiplication		
Division		

Q7 For those operations with an identity element, are there any numbers that do not have inverses?

GENERAL NOTES

This activity challenges students to analyze and understand the four fundamental arithmetic operations—addition, subtraction, multiplication, and division—from a visual perspective.

It's possible to answer any of the questions in this activity without using the arithmetic machines at all. But that defeats its purpose: Watching the machines in action as you drag pointers a and b can be fascinating. The four arithmetic operations, normally computed on a discrete, case-by-case basis, yield new insights when viewed from a continuous, motion-based perspective.

To get the most out of this activity, students should not treat the process of finding the answers as a purely mechanical one, dragging the pointers around without thinking about the meaning behind the various arrangements. Encourage students to discuss and reflect as they work.

Group Work: Dividing the class into groups will make the work more manageable and allow time for reflection. Each group can be responsible for filling out the four charts for addition, subtraction, multiplication, and division. Encourage group members to work together, observing each other's experiments and checking each other's work as they fill in the charts together, rather than having individual members work in isolation on specific charts.

Translating Between Representations: Doing this investigation requires students to translate between abstract algebraic statements and descriptions of concrete geometric behavior. First they must translate from the algebraic language in the chart ("$a \otimes b > a$ and $a \otimes b > b$") to the geometric behavior they are looking for ("The $a \otimes b$ marker is to the right of both a and b"). Once they have investigated the behavior of the model, they must then translate the geometric behavior they have observed back into algebraic language. This translation process is not easy for students, and it may be helpful to address the process explicitly, by having them fill out (perhaps as a class) the Geometric Descriptions of Algebraic Properties chart provided. Sample answers are provided following the Properties charts on the next pages.

An Alternate Approach: One possible variation to the approach suggested in the activity is to focus on one property for all four operations. For example, students might focus on the first property ($a \otimes b = 0$) and fill out the top row of each of the four charts—for addition, subtraction, multiplication, and division.

 ACTIVITY NOTES

Additional Investigations: The questions in this activity are the tip of the iceberg when it comes to exploring the arithmetic machines. Why not challenge your students to write and share some of their own questions? Here are a few to consider:

· Describe the behavior of $a \div b$ as b is dragged back and forth through 0.

· If $a \cdot b$ equals 0, does knowing the value of a allow you to determine a unique value for b?

· If $a \div b$ equals 0, does knowing the value of a allow you to determine a unique value for b?

· When is $a - b$ greater (or less) than $a + b$?

· When is $a \cdot b$ greater (or less) than $a \div b$?

· Under what circumstances does $a + b = a - b$ regardless of where you drag a?

· Under what circumstances does $a \cdot b = a \div b$ regardless of where you drag a?

INVESTIGATE

Following are sample answers for each of the four charts. The specific examples and the "When is it true?" description will vary; many correct responses are possible.

 ACTIVITY NOTES

Addition Properties

$a + b = 0$	$a = 3,$ $\quad b = -3$ $a = -2,$ $\quad b = 2$ $a = 0,$ $\quad b = 0$	The sum of two numbers is zero when the numbers are the opposites of each other (or both equal zero).	
$a + b = 1$	$a = 0,$ $\quad b = 1$ $a = 3,$ $\quad b = -2$ $a = -3,$ $\quad b = 4$	The sum of two numbers is one when the second number is one more than the opposite of the first.	
$a = b = a + b$	$a = 0,$ $\quad b = 0$	The only way two numbers can both be equal to their sum is when both numbers are zero.	
$a = a + b$	$a = 5,$ $\quad b = 0$ $a = -3,$ $\quad b = 0$ $a = 0,$ $\quad b = 0$	The sum of two numbers is equal to the first number only if the second number is zero.	
$a > 0, b > 0,$ and $a + b > 0$	$a = 1,$ $\quad b = 1$ $a = 0.5,$ $\quad b = 1$ $a = 5,$ $\quad b = 10$	When both numbers are positive, their sum is always positive.	
$a < 0, b < 0,$ and $a + b < 0$	$a = -1,$ $\quad b = -1$ $a = -0.5,$ $b = -1$ $a = -2,$ $\quad b = -3$	When both numbers are negative, their sum is always negative.	
$a + b > a$ and $a + b > b$	$a = 1,$ $\quad b = 1$ $a = 0.1,$ $\quad b = 2$ $a = 3,$ $\quad b = 4$	The sum is greater than either number provided both numbers are positive.	
$a + b$ is between a and b	$a = 1,$ $\quad b = -2$ $a = -1,$ $\quad b = 0.1$ $a = -5,$ $\quad b = 4$	If one number is positive and the other negative, the sum is between the two numbers.	
$a + b < a$ and $a + b < b$	$a = -1,$ $\quad b = -1$ $a = -1,$ $\quad b = -10$ $a = -10,$ $b = -1$	The sum is less than either number if both numbers are negative.	

Exploring Algebra 1 with The Geometer's Sketchpad
© 2012 Key Curriculum Press

Subtraction Properties

$a - b = 0$	$a = 3,$ $\quad b = 3$ $a = -2,$ $\quad b = -2$ $a = 0,$ $\quad b = 0$		The difference of two numbers is zero when the numbers are equal.
$a - b = 1$	$a = 2,$ $\quad b = 1$ $a = -1,$ $\quad b = -2$ $a = 1,$ $\quad b = 0$		The difference of two numbers is one when the first number is one more than the second.
$a = b = a - b$	$a = 0,$ $\quad b = 0$		The only way two numbers can both be equal to their difference is when both numbers are zero.
$a = a - b$	$a = 5,$ $\quad b = 0$ $a = -3,$ $\quad b = 0$ $a = 0,$ $\quad b = 0$		The difference of two numbers is equal to the first number only if the second number is zero.
$a > 0, b > 0,$ and $a - b > 0$	$a = 1,$ $\quad b = 0.5$ $a = 1.1,$ $\quad b = 1$ $a = 5,$ $\quad b = 4$		When both numbers are positive, their difference is positive if the first is greater than the second.
$a < 0, b < 0,$ and $a - b < 0$	$a = -2,$ $\quad b = -1$ $a = -1.5,$ $\quad b = -1$ $a = -4,$ $\quad b = -3$		When both numbers are negative, their difference is negative if the first number is less than the second.
$a - b > a$ and $a - b > b$	$a = -1,$ $\quad b = -1$ $a = -3,$ $\quad b = -2$ $a = 3,$ $\quad b = -4$		The difference is greater than either number as long as the second number is negative and the first is greater than twice the second.
$a - b$ is between a and b	$a = 5,$ $\quad b = 2$ $a = -9,$ $\quad b = -4$ $a = 13,$ $\quad b = 6$		If the second number is positive, the difference is between the two numbers when the first is greater than twice the second. If the second number is negative, the difference is between the two numbers when the first is less than twice the second.
$a - b < a$ and $a - b < b$	$a = 1,$ $\quad b = 1$ $a = 3,$ $\quad b = 2$ $a = 7,$ $\quad b = 4$		The difference is less than either number if the second number is positive and the first is less than twice the second.

 ACTIVITY NOTES

Multiplication Properties

$a \cdot b = 0$	$a = 0,$ $\quad b = -3$ $a = -2,$ $\quad b = 0$ $a = 0,$ $\quad b = 0$		The product of two numbers is zero when at least one of the numbers is zero.
$a \cdot b = 1$	$a = 1,$ $\quad b = 1$ $a = 3,$ $\quad b = 1/3$ $a = 1/4,$ $\quad b = 4$		The product of two numbers is one when the numbers are reciprocals.
$a = b = a \cdot b$	$a = 0,$ $\quad b = 0$ $a = 1,$ $\quad b = 1$		The only way two numbers can both be equal to their product is when both numbers are zero or both numbers are one.
$a = a \cdot b$	$a = 5,$ $\quad b = 1$ $a = -3,$ $\quad b = 1$ $a = 0,$ $\quad b = 1$		The product of two numbers is equal to the first number only if the second number is one.
$a > 0, b > 0,$ and $a \cdot b > 0$	$a = 1,$ $\quad b = 1$ $a = 0.5,$ $\quad b = 1$ $a = 5,$ $\quad b = 10$		The product of two positive numbers is positive.
$a < 0, b < 0,$ and $a \cdot b < 0$	Never		The product of two negative numbers is never negative.
$a \cdot b > a$ and $a \cdot b > b$	$a = 1.1,$ $\quad b = 1.1$ $a = -1,$ $\quad b = -2$ $a = 3,$ $\quad b = 4$		The product is greater than either number either when both numbers are negative or when both numbers are greater than one.
$a \cdot b$ is between a and b	$a = 1/2,$ $\quad b = -2$ $a = 1/2,$ $\quad b = 3$ $a = 4,$ $\quad b = 1/4$ $a = -3,$ $\quad b = 1/4$		The product is between the two numbers if one number is between zero and one and the other is either negative or greater than one.
$a \cdot b < a$ and $a \cdot b < b$	$a = 0.5,$ $\quad b = 0.2$ $a = -2,$ $\quad b = 1.5$ $a = 2,$ $\quad b = -0.1$		The product is less than either number if both numbers are between zero and one, or if one number is negative and the other is greater than one.

Exploring Algebra 1 with The Geometer's Sketchpad
© 2012 Key Curriculum Press

Division Properties

$a \div b = 0$	$a = 0,\quad b = -3$ $a = 0,\quad b = 2$ $a = 0,\quad b = 1/2$	The quotient of two numbers is zero when the numerator is zero and the denominator is not zero.
$a \div b = 1$	$a = 1,\quad b = 1$ $a = -2,\quad b = -2$ $a = 4,\quad b = 4$	The quotient of two numbers is one when the numbers are equal but not zero.
$a = b = a \div b$	$a = 1,\quad b = 1$	The only way two numbers can both be equal to their quotient is when both numbers are one.
$a = a \div b$	$a = 5,\quad b = 1$ $a = -3,\quad b = 1$ $a = 0,\quad b = 1$	The quotient of two numbers is equal to the numerator only if the denominator is one.
$a > 0, b > 0,$ and $a \div b > 0$	$a = 1,\quad b = 1$ $a = 0.5,\quad b = 1$ $a = 5,\quad b = 10$	The quotient of two positive numbers is always positive.
$a < 0, b < 0,$ and $a \div b < 0$	Never	The quotient of two negative numbers is never negative.
$a \div b > a$ and $a \div b > b$	$a = -1,\quad b = -2$ $a = 0.2,\quad b = 0.5$ $a = 0.2,\quad b = 0.1$	The quotient is greater than either number if both numbers are negative, or if the denominator is between zero and one and the numerator is greater than the square of the denominator.
$a \div b$ is between a and b	$a = 5,\quad b = 2$ $a = 1/8,\quad b = 1/2$ $a = 8,\quad b = -4$	The quotient is between the two numbers if the denominator is greater than one and the numerator is greater than the square of the denominator, or if the denominator is less than one and the numerator is between zero and the square of the denominator.
$a \div b < a$ and $a \div b < b$	$a = 5,\quad b = -2$ $a = -1,\quad b = 1/2$ $a = 3,\quad b = 2$	The quotient is less than either number if (a) the denominator is less than zero and the numerator is greater than the square of the denominator, or (b) the denominator is between zero and one and the numerator is less than zero, or (c) the denominator is greater than one and the numerator is between zero and the square of the denominator.

 ACTIVITY NOTES

Geometric Descriptions of Algebraic Properties

If you have students fill in this chart, here are sample geometric descriptions of the algebraic properties:

Row	Algebraic Property	Geometric Description
1	$a \otimes b = 0$	The $a \otimes b$ marker points at 0.
2	$a \otimes b = 1$	The $a \otimes b$ marker points at 1.
3	$a = b = a \otimes b$	The a, b, and $a \otimes b$ markers all point to the same location.
4	$a = a \otimes b$	The a and $a \otimes b$ markers point to the same location.
5	$a > 0$, $b > 0$, and $a \otimes b > 0$	The a, b, and $a \otimes b$ markers are all to the right of 0.
6	$a < 0$, $b < 0$, and $a \otimes b < 0$	The a, b, and $a \otimes b$ markers are all to the left of 0.
7	$a \otimes b > a$ and $a \otimes b > b$	The $a \otimes b$ marker is to the right of both the a and the b markers.
8	$a \otimes b$ is between a and b	The $a \otimes b$ marker is between a and b. (Note that the algebraic description given here isn't very algebraic. The real algebraic description would be $a < a \otimes b < b$ or $b < a \otimes b < a$.)
9	$a \otimes b < a$ and $a \otimes b < b$	The $a \otimes b$ marker is to the left of both a and b.

Exploring Algebra 1 with The Geometer's Sketchpad
© 2012 Key Curriculum Press

Exploring Properties of Operations

In this activity you'll use Sketchpad arithmetic machines to investigate the properties of the four fundamental arithmetic operations.

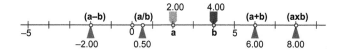

INVESTIGATE

1. Open **Operation Properties.gsp.**

2. On the "Addition" page, use the **Add** custom tool to construct a marker showing the sum of *a* and *b*.

3. On a copy of the Operation Properties chart, make the title of the chart "Operation Properties for Addition," and fill in the blanks in the "Property" column with addition signs.

4. For each row of the chart, experiment by dragging *a* and *b* to determine whether that row's property is possible. If the property is not possible, write "Never" and explain why in the "When is it true?" column. If possible, fill in three examples, and write a sentence in the "When is it true?" column describing the conditions that must be met for the description to be true. Be specific about the values that the two numbers can have.

5. When you finish with one operation, go to the next page of the sketch, and use the appropriate custom tool to construct a result marker on that page. Then fill out the Operation Properties chart for that operation.

Here's an example of what you might write in the first row of the Addition chart:

Property	Examples	When is it true?
$a + b = 0$	$a = 5,\quad b = -5$ $a = -3,\quad b = 3$ $a = 0,\quad b = 0$	The sum of two numbers is zero when the numbers are the opposites of each other (or both equal zero).

PRESENT

You can use the Help system to learn how to make and use Movement, Animation, and Hide/Show buttons.

On the "Presentation" page of the document, choose one particular property of one of the operations, and create a presentation sketch that uses Movement, Animation, or Hide/Show buttons to demonstrate the circumstances in which the property is true.

Operation Properties for _____

Property	Examples		When is it true?
$a \square b = 0$	$a =$	$b =$	
	$a =$	$b =$	
	$a =$	$b =$	
$a \square b = 1$	$a =$	$b =$	
	$a =$	$b =$	
	$a =$	$b =$	
$a = b = a \square b$	$a =$	$b =$	
	$a =$	$b =$	
	$a =$	$b =$	
$a = a \square b$	$a =$	$b =$	
	$a =$	$b =$	
	$a =$	$b =$	
$a > 0, b > 0$, and $a \square b > 0$	$a =$	$b =$	
	$a =$	$b =$	
	$a =$	$b =$	
$a < 0, b < 0$, and $a \square b < 0$	$a =$	$b =$	
	$a =$	$b =$	
	$a =$	$b =$	
$a \square b > a$ and $a \square b > b$	$a =$	$b =$	
	$a =$	$b =$	
	$a =$	$b =$	
$a \square b$ is between a and b	$a =$	$b =$	
	$a =$	$b =$	
	$a =$	$b =$	
$a \square b < a$ and $a \square b < b$	$a =$	$b =$	
	$a =$	$b =$	
	$a =$	$b =$	

Exploring Algebra 1 with The Geometer's Sketchpad
© 2012 Key Curriculum Press

Geometric Descriptions of Algebraic Properties

You can use this chart to help you figure out what geometric features to look for when you investigate a particular algebraic property. For the algebraic property in each row of the table, write a sentence describing the corresponding geometric behavior of the markers for *a, b,* and $a \otimes b$. Row 7 is filled in as an example. ($\otimes$ stands for $+, -, \cdot,$ or $\div.$)

Row	Algebraic Property	Geometric Description
1	$a \otimes b = 0$	
2	$a \otimes b = 1$	
3	$a = b = a \otimes b$	
4	$a = a \otimes b$	
5	$a > 0, b > 0,$ and $a \otimes b > 0$	
6	$a < 0, b < 0,$ and $a \otimes b < 0$	
7	$a \otimes b > a$ and $a \otimes b > b$	The $a \otimes b$ marker is to the right of both the *a* and the *b* marker.
8	$a \otimes b$ is between *a* and *b*	
9	$a \otimes b < a$ and $a \otimes b < b$	

2

Ratios and Exponents

Ratio and Proportion

Students manipulate two similar rectangles and study the proportions formed from the ratios of their sides.

Proportions in Similar Triangles

Students manipulate two similar triangles, measure their angles and sides, and draw conclusions about the ratios of the sides. Students then use similar triangles to solve problems involving missing sides.

Rates and Ratios

Students use a dynamic model of a machine to explore rates and proportions. They examine their assumptions about such problems, and then use proportions to explore rate problems.

The Golden Rectangle and Ratio

Students use a ratio to describe the shape of a rectangle. They construct golden rectangles and a golden spiral and examine the ratios of the rectangles in their constructions.

Fractals and Ratios

Students use repeated constructions and then iterated constructions to build fractals. They investigate the property of self-similarity and measure the ratio of similarity.

Length of the Koch Curve

Students build a Koch curve and investigate its properties. They use fractions and exponents to calculate its length at various depths and make conjectures about its overall length.

The Chaos Game

Students explore chaos by creating a recursive point mapping using a ratio and a random component. They then vary the ratio and observe changes in the patterns.

Exponents

Students use a dynamic model to explore repeated multiplication and properties of exponents.

Zero and Negative Exponents

Students create a sequence of bars to compare various integer powers of a given base. From the pattern formed, they learn to interpret zero and negative exponents.

Ratio and Proportion

INVESTIGATE

Q1 The ratio shows a fraction representing the height of the rectangle divided by its width.

Q2 As you drag point *Adjust,* the yellow rectangle changes size and shape. It remains the same shape as the blue rectangle, and the same relative size.

Q3 As you drag point *Multiplier,* the size (but not the shape) of the yellow rectangle changes. When the multiplier is 1.000, the two rectangles are the same size.

Q4 You can drag point *Multiplier* until its value is 0.500. You can also look at the height and width measurements.

Q5 To make the yellow rectangle twice as big, drag point *Multiplier* until its value is 2.000. The height of the yellow rectangle is 20.00, and the width is 30.00.

Q6 To make the yellow rectangle's height 6 when the blue one's height is 8, you must use a multiplier of 0.750. The resulting width of the yellow rectangle is 7.50.

Q7 $w = 15$, $h = 5$, $x = 0.34$, and $m = 17.78$.

EXPLORE MORE

Q8 Answers will vary. One easy way to detect incorrect proportions is to set the multiplier to 1.00 and inspect the resulting fractions. Some of the incorrect ones (such as the one on the left below) can easily be identified.

$$\frac{10.00}{15.00} = \frac{15.00}{10.00} \qquad \frac{10.00}{15.00} = \frac{10.00}{15.00}$$

This method allows students to eliminate choices *b, c, e,* and *g.*

Another method is to adjust the multiplier so that the two ratios that make up a particular proportion have equal numerators, and then inspect the denominators. To test proportion *d,* you could manipulate the multiplier to be 1.50, causing *d* to appear as shown here. This proportion is obviously false because the two numerators match but the denominators don't.

$$\frac{15.00}{22.50} = \frac{15.00}{10.00}$$

The correct proportions are *a, f,* and *h.*

Q9 Here are the correct proportions expressed numerically and symbolically:

$$(a)\quad \frac{10}{15} = \frac{20}{30} \qquad \frac{h_1}{w_1} = \frac{h_2}{w_2}$$

$$(f)\quad \frac{20}{30} = \frac{10}{15} \qquad \frac{h_2}{w_2} = \frac{h_1}{w_1}$$

$$(h)\quad \frac{30}{20} = \frac{15}{10} \qquad \frac{w_2}{h_2} = \frac{w_1}{h_1}$$

You can use the square on page 4 to check the results on page 2. If you press the *Show Captions* button, you can edit the captions to change them to the corresponding numbers. The numbers will be displayed at the corners of the square, making it easy to check the proportions by transforming the square.

Q10 Here are the eight ratios that can be arranged in pairs:

$$\frac{h_1}{w_1} = \frac{h_2}{w_2} \qquad \frac{h_1}{h_2} = \frac{w_1}{w_2} \qquad \frac{w_1}{h_1} = \frac{w_2}{h_2} \qquad \frac{h_2}{h_1} = \frac{w_2}{w_1}$$

Here are the four ratios that cannot be arranged in pairs:

$$\frac{h_1}{w_2} \qquad \frac{w_1}{h_2} \qquad \frac{w_2}{h_1} \qquad \frac{h_2}{w_1}$$

WHOLE-CLASS PRESENTATION

In this presentation students will observe details of a model of a proportion using similar rectangles, manipulate the proportion by manipulating the model, use the model to solve various proportion problems, and investigate how any proportion can be written as an equation in eight different (but symmetrical) ways.

Begin by exploring the ratio of side lengths in a rectangle.

1. Open the sketch **Proportion Present.gsp** and drag point *Adjust.*

Q1 Ask, "What does dragging the point do to the rectangle?" After several answers, ask what two things about the rectangle are changed by moving point *Adjust,* and get a student to summarize that the point changes both the size and the shape of the rectangle.

2. Press the button labeled *Show Shape Buttons.* Press the shape buttons in turn to illustrate several different shapes.

3. Put the rectangle back into its original shape.

4. Press the *Show Ratio* button.

Q2 Ask, "What does the ratio represent?" Don't expect or impose any specific answers, but explore this question by changing the rectangle's shape.

Q3 Use either point *Adjust* or the shape buttons to make the rectangle tall and thin, and ask, "Is the ratio now a large number or a small one?"

Q4 Make the rectangle short and squat, and ask whether this ratio is a large number or a small one.

Q5 Ask, "What do you think will happen to the ratio if I press the *Square* button?" Press the button to test their conjectures.

Now add a second rectangle and look at its ratio.

5. Press the *Show Yellow Rectangle* button and drag point *Adjust*.

Q6 Ask, "What relationships do you see between the two rectangles as I drag *Adjust*?" Try to elicit student observations about both the relative shapes and the relative sizes.

6. Press the *Show Multiplier* button and drag point *Multiplier*.

Q7 Ask, "How does point *Multiplier* affect the relationship between the rectangles? How does it affect the relative shapes? How does it affect the relative sizes? What do you notice when the multiplier is 1.00? When it's 2.00?"

7. Set the multiplier back to 0.5. Press the *Show Second Ratio* button.

Q8 There's an equal sign between the ratios; ask, "Are the two ratios really equal?" Point out that the fact that they are equal means that the equation shown is a *proportion*.

8. Change the shape of the rectangles by dragging point *Adjust* or by pressing the shape buttons.

Q9 Ask, "Are the ratios still equal?" Ask, "How do you think the value of the multiplier is related to the numbers making up the two ratios?" Look at differently shaped rectangles to confirm students' conjectures.

Next use the rectangles to solve some proportion problems.

9. Go to page 2 and press the *Problem 1* button to display a proportion problem.

10. Have a student manipulate point *Adjust* so that the left side of the bottom proportion matches the left side of the problem. (When the point is close to the correct position, you can use the *Ratio of Integers* button to move it to the exact integer position.)

11. Have the student drag point *Multiplier* until the height of the yellow rectangle matches the number on the right side of the problem.

Q10 Ask, "Can you find the missing number in the problem by looking at the bottom proportion? What is the missing number?"

12. Problems 2, 3, and 4 on this page present additional challenges. To solve problems 3 and 4, show the size controls and press the smaller or larger buttons to make the scale of the rectangles appropriate for the numbers in the problem.

Explore how a particular proportion can be written in several different ways.

13. Go to page 3 and use the Calculator to compute various ratios. Make sure that students see that there are a number of different proportions that they can write using the same set of numbers.

14. The square on page 4 presents an animation of the symmetry involved in the eight different ways of expressing the same relationship. Experiment with it to see the various ways of expressing the same proportion.

Exploring Algebra 1 with The Geometer's Sketchpad

Ratio and Proportion

In this activity you will use a visual model of ratios and proportions in similar rectangles to solve problems involving proportions.

INVESTIGATE

1. Open **Proportion.gsp.** Drag point *Adjust* to see what it does.

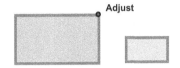

2. A proportion requires two ratios. Show the first one by pressing the *Show Ratio* button. Then drag point *Adjust* again.

Q1 What does this ratio show?

3. The second ratio will come from a second rectangle. Press the *Show Yellow Rectangle* button, and then drag *Adjust* again.

Q2 What happens to the yellow rectangle as you drag *Adjust*? What determines the shape and size of the yellow rectangle?

4. Press the *Show Multiplier* button and then drag point *Multiplier*.

Q3 What happens to the yellow rectangle?

Q4 How can you make each side of the yellow rectangle half as big as the corresponding side of the blue one?

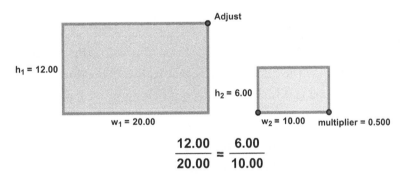

$$\frac{12.00}{20.00} = \frac{6.00}{10.00}$$

5. To complete the proportion, press the *Show Second Ratio* button, and again drag points *Adjust* and *Multiplier*. Observe the effects on the proportion when you drag these two points.

The *Ratio of Integers* button can help you get the numbers exact.

6. Adjust the blue rectangle so its height-to-width ratio is 10.00/15.00. Press the *Set Multiplier to 1* button, and then make the yellow rectangle twice as big as the blue one horizontally and vertically.

Q5 What did you do to make the sides of the yellow rectangle twice as big as the sides of the blue one? What are the height and width of the yellow rectangle?

7. Adjust the blue rectangle so that h_1/w_1 is 8.00/10.00, and adjust the multiplier so that the yellow rectangle's height (h_2) is 6.00.

Q6 What multiplier did you use to do this? What is the resulting width (w_2) of the yellow rectangle?

Q7 In step 7, you actually solved the proportion $\frac{8.00}{10.00} = \frac{6.00}{w_2}$. Use the rectangles to solve the following proportions:

Show the size controls if you need to change the size of the rectangles.

$$\frac{8}{10} = \frac{12}{w} \qquad \frac{2.00}{3.00} = \frac{h}{7.50} \qquad \frac{x}{0.30} = \frac{1.70}{1.50} \qquad \frac{32}{m} = \frac{81}{45}$$

EXPLORE MORE

Go to page 2. This page lists eight different proportions using the height and width of the two rectangles. Some of the proportions are correct mathematically, but some are wrong. Determine which ratios are correct by manipulating the rectangles.

You can identify some incorrect proportions by making the rectangles square, and you can identify others by setting the multiplier to 1.

Q8 How can you tell which proportions are correct by manipulating the rectangles?

Q9 For each correct proportion, write down its letter, then the proportion as it appears using numbers, and finally the proportion using the symbols h_1, w_1, h_2, and w_2 in place of the numbers. For instance, proportion (a) is $\frac{10}{15} = \frac{20}{30}$, so you would write

$$(a) \quad \frac{10}{15} = \frac{20}{30} \qquad \frac{h_1}{w_1} = \frac{h_2}{w_2}$$

Choose **Number | Calculate** to use Sketchpad's Calculator. Click a measurement in the sketch to enter its value into the Calculator.

Q10 On page 3, calculate each of the 12 possible ratios involving h_1, w_1, h_2, and w_2. Arrange the ratios in pairs that have equal values. Write a proportion for each such pair. Which ratios do not belong to such pairs?

Go to page 4. This page shows the four variables from the rectangles at the four corners of a square. By reflecting the square across one of its axes of symmetry or by rotating by a multiple of 90°, you can generate all possible correct proportions. Use the square to check your answers for the proportions from pages 2 and 3.

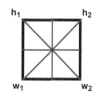

Exploring Algebra 1 with The Geometer's Sketchpad
© 2012 Key Curriculum Press

Proportions in Similar Triangles

INVESTIGATE

Q1 As you drag points *D* and *E*, the angles of the blue triangle stay the same, but its size, position, and orientation change.

Q2 The *Present Similarity* button shows an animation that suggests the two triangles are the same shape.

Q3 The triangles match up when point *D* is on top of *A* and *E* is on top of *B*.

Q4 When you make ∠*BAC* in the yellow triangle very small, ∠*EDF* in the blue triangle also becomes small.

Q5 When you drag point *C* far away from points *A* and *B*, point *F* moves far away from points *D* and *E*.

Q6 Angle *DEF* in the blue triangle corresponds to ∠*ABC* in the yellow triangle, and ∠*EFD* corresponds to ∠*BCA*. Angle *CAB* in the yellow triangle corresponds to ∠*FDE* in the blue triangle.

Q7 The corresponding angles of similar triangles are equal.

Q8 The ratios of the three pairs of corresponding sides of similar triangles are equal.

Q9 Students should write these three proportions or their equivalents:

$$\frac{m\overline{AB}}{m\overline{ED}} = \frac{m\overline{CA}}{m\overline{DF}} \qquad \frac{m\overline{AB}}{m\overline{ED}} = \frac{m\overline{BC}}{m\overline{EF}} \qquad \frac{m\overline{BC}}{m\overline{EF}} = \frac{m\overline{CA}}{m\overline{DF}}$$

Q10 Corresponding angles of similar triangles are equal, and the ratios of corresponding sides are equal.

Q11 $m\overline{EF} = 6.7$ cm. This result is rounded off to the nearest tenth, as are all the distances in this activity.

Q12 Because of rounding, some students may end up with slightly different answers such as $m\overline{EF} = 6.6$ cm.

Q13 $m\overline{AB} = 2.6$ cm. The proportion is

$$\frac{m\overline{AB}}{m\overline{ED}} = \frac{m\overline{CA}}{m\overline{DF}}$$

EXPLORE MORE

Q14 Although this problem is conceptually the same as the similar triangles in the activity, students may have trouble getting it. There are still

two triangles. One triangle appears in the scale drawing, with sides that students should measure in cm. The other straddles the banks of the river itself, with one side of 500 m and another side that is the unknown length of the bridge. Students need to set up a proportion using the corresponding sides of these triangles. The bridge must be 888 m in length.

WHOLE-CLASS PRESENTATION

In this presentation students will observe the behavior of similar triangles, compare measurements of their sides and angles, find ratios that are equal, and use the ratios to write proportions. Students will then use the proportions to solve several different problems involving similar triangles.

To present this activity to the entire class, follow the Presenter Notes and use the sketch **Similar Triangles Present.gsp.**

Exploring Algebra 1 with The Geometer's Sketchpad
© 2012 Key Curriculum Press

Proportions in Similar Triangles

1. Open **Similar Triangles Present.gsp.** Drag points *D* and *E*.

Q1 Ask, "What do you notice about the triangles as I drag points *D* and *E*?" [The answers should include the fact that the position, orientation, and size change, and should also include the observation that the shape doesn't change or that the sizes of the angles don't change.]

You may want to use the Present in Reverse button after the Present Similarity button.

2. Press the *Present Similarity* button. As the animation takes place, point out the three stages of the animation: translation (change of position); rotation (change of orientation); and dilation (change of size).

3. Show the vertices and drag points *A*, *B*, and *C*.

Q2 As you drag points *A*, *B*, and *C*, ask students to identify which vertex of the blue triangle corresponds to each vertex of the yellow one.

When the triangles are aligned, you can drag the blue one so it lies precisely on top of the yellow one.

4. Check the answers by clicking the *Align Triangles* button. With the triangles aligned, press the *Show Angle Measurements* button and the *Show Distance Measurements* button.

Q3 Ask students what they notice about the measurements. [Both sets of measurements are exactly equal between the two triangles.]

Q4 Drag the vertices and ask what students notice. [The angles remain equal, but the distances don't.]

Q5 Press *Show Ratios* and ask students what they observe about the ratios. [All three are equal.] Drag vertices to change relative sizes; students should observe that the ratios change, but all three remain equal to each other.

Tell students it's not fair to interchange the left and right sides of a proportion and count it as different.

You can also use the buttons in the sketch to adjust the distances.

Q6 Point out that proportions are made up of two equal ratios, and ask students to write down as many different proportions as they can from the three ratios. Press the *Show Proportions* button to verify the answers.

Q7 Go to page 2 of the sketch, and ask students to write and solve a proportion for the problem that appears. Then show the distance measurements and drag point *B* until *AB* = 3.0 cm, drag point *C* until *BC* = 4.0 cm, and drag point *E* until *ED* = 5.0 cm. Have students compare their results with distance *EF*.

5. Use a similar method to solve the problem listed on page 3 of the sketch.

6. On the Explore More page of the sketch, show the hint and let students read it.

7. Use the buttons to show a triangle, to show the distance measurements, and to make the triangle similar to triangle *ABC* on the map.

Q8 Ask students to set up a proportion using the two triangles and to find the length of the bridge. Use the buttons in the sketch to check their answers.

Proportions in Similar Triangles

Two geometric figures are *similar* if they have the same shape but not necessarily the same size. In this activity you'll investigate the properties of similar triangles and use proportions to find the missing sides of a pair of triangles.

INVESTIGATE

1. Open **Similar Triangles.gsp.**

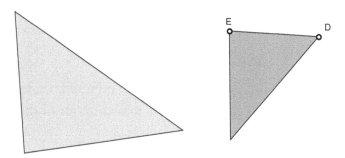

2. Drag points *D* and *E* to change the blue triangle.

Q1 As you drag the points, what changes and what stays the same?

Q2 Press the *Present Similarity* button. What does this seem to show about the two triangles?

3. To change the shape of the yellow triangle, you must first show its vertices. Press the *Show Vertices* button.

4. Drag the vertices of each triangle. Notice what changes and what stays the same.

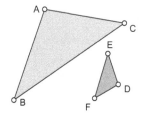

Q3 Drag points *D* and *E* until △*DEF* exactly overlays △*ABC*. Where do you have to place points *D* and *E* to make the triangles match up?

5. Drag the triangles apart again.

Q4 Make ∠*CAB* in the yellow triangle very small. What happens in the blue triangle?

Q5 Drag point *C* far away from points *A* and *B*. What happens in the blue triangle?

6. Press the *Show Angle Measurements* button.

Q6 Drag each vertex to see what happens to the angle measurements. Which angle in the blue triangle corresponds to ∠*ABC* in the yellow triangle? Which angle corresponds to ∠*BCA*? Which angle in the yellow triangle corresponds to ∠*FDE*?

Q7 What can you conclude about the corresponding angles of similar triangles?

Exploring Algebra 1 with The Geometer's Sketchpad
© 2012 Key Curriculum Press

For GSP5

To measure the length of a side, select it and choose **Measure | Length.** You may select all the sides and measure the lengths all at one time.

7. Measure all the sides of each triangle.

8. Calculate the ratio of one pair of corresponding sides. To do so, choose **Number | Calculate,** click in the sketch on a length measurement from the yellow triangle, click the division symbol in the Calculator's keypad, and click the corresponding length measurement from the blue triangle.

$$m\,\overline{AB} = 6.4 \text{ cm} \qquad m\,\overline{ED} = 4.2 \text{ cm} \qquad \frac{m\,\overline{AB}}{m\,\overline{ED}} = 1.53$$
$$m\,\overline{BC} = 7.6 \text{ cm} \qquad m\,\overline{EF} = 4.9 \text{ cm}$$
$$m\,\overline{CA} = 9.6 \text{ cm} \qquad m\,\overline{DF} = 6.3 \text{ cm}$$

9. Calculate the ratios of the other two pairs of corresponding sides.

Q8 What can you conclude about the ratios of the corresponding sides of these triangles?

Q9 Complete the following proportion using the ratios you calculated in step 8 and step 9. Then write two other proportions using these ratios.

$$\frac{m\overline{AB}}{} = \frac{}{m\overline{DF}}$$

Q10 Using your answers to Q7 and Q8, write a summary of the properties of similar triangles.

Q11 If $m\overline{AB} = 3.0$ cm, $m\overline{BC} = 4.0$ cm, and $m\overline{ED} = 5.0$ cm, use the triangles to find $m\overline{EF}$. Set up a proportion to check your answer.

Q12 Compare your answer to Q11 with other students' answers. Then compare your sketches. Explain any similarities or differences.

Q13 If $m\overline{CA} = 4.5$ cm, $m\overline{ED} = 2.0$ cm, and $m\overline{DF} = 3.5$ cm, use the triangles to find $m\overline{AB}$. Set up a proportion to check your answer.

EXPLORE MORE

Q14 Page 2 of the sketch shows a scale drawing of a river and the location of a bridge that must be built between point *A* and point *B*. There is no easy way to measure directly the distance between the two points on opposite sides of the river. Use similar triangles and proportions to find the required length of the bridge.

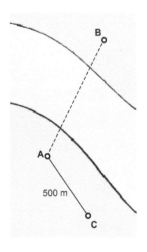

Rates and Ratios

 **ACTIVITY NOTES**

Some textbooks describe "cross-multiplying" of proportions, and some avoid the technique. Q11 asks students to justify this technique. This question is optional, and you can tell students to skip it if you prefer.

EXAMINE YOUR ASSUMPTIONS

The model used in this activity is one way to see equal ratios. The visual model is designed to illustrate repeating a rate—every 4 seconds, you will see 7 more noodles. The model also marks off, with the red segments, groups of 7 noodles and 4 seconds, so that you can see that you have made 8 copies of the original ratio. The discrete nature of the machine makes it easier for students to see these copies. If students are actually counting noodles and seconds, they must count the colored intervals between the vertical line segments, not the segments themselves.

Students can change the rate at which their machine works by changing the rate parameters at the top of the sketch: the number of noodles and the number of seconds.

Q1 If you are buying 3 boxes of detergent, and each box has the same number of ounces and costs the same, then you can repeat the ratio in this way. However, students may be familiar with buying in bulk for a lower cost.

Q2 The runner may be able to keep up that same speed while running 10 times as far, but then again, she may not.

Q3 The machine produces 7 noodles in one turn of the crank, which takes 4 seconds. The ratio is

$$\frac{7 \text{ noodles}}{4 \text{ seconds}}$$

Q4 After two turns of the crank, there are 14 noodles, and the machine has been running for 8 seconds.

Q5 There are 8 copies of 4 seconds in 32 seconds, so there are 8 copies of 7 noodles, making 56 noodles.

Q6 The pasta machine produces the same number of noodles every 4 seconds and doesn't slow down or speed up over time. Some students will argue that this means it runs at a constant rate. Other students may point out that the machine doesn't produce any finished noodles until the end of the 4 seconds, so it's not really running at a constant

rate—and that at 2 seconds, it has not produced even a single finished noodle. Both arguments have merit, and this question can lead to a very enlightening class discussion on what is meant by *constant rate*, and on the difference between discrete behavior (exemplified by the pasta machine on page 1) and continuous behavior (exemplified by the machine on page 2).

RATES AND PROPORTIONS

No matter how a proportion is solved, students should have a sense of the meaning of their calculations. Solving by finding the number of copies as in Q7 focuses students on the idea that they are repeating a rate.

Q7 You must repeat the 4-second ratio 7.5 times. You can see this in the model by thinking of 4 as a unit; there are 7.5 groups of 4 seconds in 30 seconds. The red segments shown in the model may help. Since you have 7.5 groups of 4 seconds, you will also have 7.5 groups of 7 noodles, or 52.5 noodles.

Q8 Calculate 30 seconds divided by 4 seconds times 7 noodles. Dividing 30 by 4 means 7.5 groups of 4 seconds. Since the rate is constant, there will be 7.5 groups of 7 noodles. Multiply 7.5 by 7.

Q9 Divide to find the number of times the ratio will be repeated. Multiply this number by the other quantity you are repeating.

Q10 Since 105 noodles equals $2y$ noodles, the value of y is 52.5. The two rates (y noodles in 30 seconds and 7 noodles in 4 seconds) are equal, so repeating both will result in ratios that are still equal. Whichever equation you use to solve for y, the value of y is the same.

Q11 Take the present question as an example. You would multiply 30 by 7, and 4 by y. The meaning of these steps is clearer if you write

$$\frac{30(7 \text{ noodles})}{30(4 \text{ seconds})} = \frac{4(y \text{ noodles})}{4(30 \text{ seconds})}$$

This means that you have repeated the ratio on the left 30 times and the one on the right 4 times. Since each ratio gives the number of noodles made in 120 seconds, the numerators must be equal. Since $4y$ equals 210, y equals 52.5.

EXPLORE MORE

Q12 To calculate a unit rate equivalent to the given rate of 7 noodles in 4 seconds, divide both the numerator and denominator by 4. The unit rate is 1.75 noodles per second. Knowing how many noodles are made in a second allows you to multiply by 30 to find how many noodles were made in 30 seconds. As before, you are repeating a constant rate.

Q13 Divide to find a unit rate; multiply this unit rate by the number of units to which you are repeating the rate.

WHOLE-CLASS PRESENTATION

In this presentation students will observe a machine running at a certain rate and use a proportion to figure out how much it will produce during a certain period of time. Students will discuss what a constant rate means, why it's required to use a proportion, and how continuous and discrete processes differ.

Start with a Sketchpad model of a pasta machine producing noodles at a certain rate.

1. Open **Rates and Ratios.gsp.** Press the *Turn the Crank* button.

Q1 Ask, "How many noodles did the machine produce in one turn of the crank? How long did it take?" (Students can use the gold rectangles to count the noodles, and the blue ones to count the seconds.)

Q2 Ask students to write the ratio of noodles to seconds as a fraction.

Q3 Turn the crank again and ask students how many noodles there are now, and how many seconds the machine has been running.

2. Turn the crank 6 more times, so the machine has run for a total of 32 seconds.

Q4 Ask students to determine how many noodles the machine has made without counting them. Ask how they got their answers. Express the student explanations as proportions.

Discuss what a *constant rate* means.

Q6 Ask whether the pasta machine runs at a constant rate. This question does not have a clear-cut answer, and should lead to an interesting discussion about the meaning of constant rate and about discrete and continuous processes.

A slightly different machine can help students understand the idea of constant rate.

3. Go to page 2 and press the *Turn the Crank Once* button.

Q7 Ask students how this machine differs from the machine on page 1. Turn the crank again to help them observe the differences.

Q8 Ask students to figure out how many noodles this machine can produce in 30 seconds. They should do their calculation by setting up a proportion. Once they have their answers, check the answers by dragging point *time* to 30 seconds.

Have students summarize the results.

Q9 Ask students to describe the method they used to find the number of noodles so that another person could use the method to solve any proportion.

Q10 Ask students to explain why using proportions works only if the machine runs at a constant rate.

Rates and Ratios

You buy 52 ounces of detergent for 9 dollars. You run 100 meters in 15 seconds. You use 5 gallons of gas to drive 150 miles. Each of these pairs of numbers is an example of a *ratio*.

You can use ratios to answer questions like: If my pasta machine makes 7 noodles in 4 seconds, how many noodles will it make in 30 seconds? To answer such questions, you have to make some assumptions. In this activity, you will examine what those assumptions are and use them to answer questions like the ones posed here.

EXAMINE YOUR ASSUMPTIONS

This ratio is sometimes written as 52:9, and sometimes as $\frac{52}{9}$.

Suppose you can buy 52 ounces of detergent for 9 dollars. If you triple both the values in that ratio, you'll get 156 ounces for 27 dollars. The second ratio is considered equal to the first, because it consists of 3 copies of the original ratio.

Any pair of numbers, like the ones mentioned above, can be considered a ratio. Not all ratios, however, can be applied in the same way.

Q1 In your experience, if 52 ounces of detergent cost 9 dollars, will 156 ounces cost 27 dollars? Why or why not?

Q2 In your experience, if a person ran 100 meters in 15 seconds, will that person run 1000 meters in 150 seconds? Why or why not?

A *constant rate* is a rate that stays the same over time.

In both examples above, you can create a new ratio equal to the original ratio. Whether the new ratio is meaningful depends upon whether your ratio represents a *constant rate* (a rate that stays the same over time).

Imagine that each turn of the crank on your pasta machine takes a certain amount of time and produces a certain number of noodles.

1. Open **Rates and Ratios.gsp.**

2. To run the machine, press the *Turn the Crank* button.

Use the gold rectangles to count the noodles, and the blue ones to count the seconds.

Q3 How many noodles does the machine produce in one turn of the crank? How long does it take? Write the ratio of noodles to seconds as a fraction.

Q4 Turn the crank again. How many noodles do you have now? How many seconds has the machine been running?

Q5 Keep turning the crank until the machine has run for 32 seconds total. Without counting them, determine how many noodles the machine has made. How did you figure this out without counting?

Q6 Does the pasta machine run at a constant rate? Explain.

Exploring Algebra 1 with The Geometer's Sketchpad
© 2012 Key Curriculum Press

RATES AND PROPORTIONS

A *proportion* is a statement that two ratios are equal. You used the pasta machine to show the rate at which the machine runs in the form of two different ratios:

$$\frac{7 \text{ noodles}}{4 \text{ seconds}} = \frac{56 \text{ noodles}}{32 \text{ seconds}}$$

The second ratio extends the basic rate (7 noodles in 4 seconds) 8 times.

3. Go to page 2. This model of the pasta machine runs at the same rate but produces noodles continuously rather than in batches.

Since the machine runs at a constant rate, the rate for 4 seconds is the same as the rate for 30 seconds, and you can use the proportion

$$\frac{7 \text{ noodles}}{4 \text{ seconds}} = \frac{y \text{ noodles}}{30 \text{ seconds}}$$

to determine how many noodles the machine can make in 30 seconds.

Model this by dragging the time slider—but don't count your noodles!

Q7 How many times do you need to repeat the rate "7 noodles in 4 seconds" to reach 30 seconds? Explain how to use this value to determine the value of *y*.

Q8 Describe a series of calculations that produces the value of *y* using the quantities 7 noodles, 4 seconds, and 30 seconds. Explain why this method works.

Q9 Describe the method you used in the previous question so that another person could use the method to solve any proportion.

Q10 If you repeat the ratio "7 noodles in 4 seconds" 15 times, you get 15 · 7 noodles in 15 · 4 seconds. That's 105 noodles in 60 seconds. If you repeat the ratio "*y* noodles in 30 seconds" 2 times, that's 2*y* noodles in 60 seconds. What is the value of *y*? Should it be the same as the value you found in Q7? Why?

Q11 If you've heard about "cross-multiplying" as a way to solve a proportion, explain why it works using the idea of "repeating" both ratios in a proportion.

EXPLORE MORE

Q12 Reduce the number of seconds to exactly 1 by dragging point *time*. How many noodles does the machine make in 1 second? How did you calculate this? Does multiplying this value by 30 give you the number of noodles made in 30 seconds? Why?

Q13 Describe the method you used in the previous question so that another person could use the method to solve any proportion.

The Golden Rectangle and Ratio

SKETCH AND INVESTIGATE

If students are new to the tools, identify the **Segment** tool, the **Point** tool, the **Text** tool, and the **Custom** tool icon. Students must use the **Text** tool in several places in the activity in order to show, hide, or change point labels.

2. Students can construct the perpendiculars one at a time by selecting segment *AB* and one point before choosing the command, or they can construct both at once by selecting segment *AB* and both points before choosing the command.

4. To construct intersection *D*, students can use the **Point** tool or the **Arrow** tool, or they can select the two lines and choose **Construct | Intersection**.

6. This step assumes that students can find **Construct | Quadrilateral Interior** once they have selected the four points.

Q1 Drag point *C* to change the shape of the rectangle. Dragging points *A* and *B* changes the size of the rectangle but not its shape.

Q2 Answers will vary, depending on the shape of the original rectangle.

Q3 Students will not be able to get the two ratios exactly equal due to limitations of dragging points on the screen, but they should get an answer slightly greater than 1.6. When the ratios are equal, they are approximately 1.618. This is the golden ratio (φ).

Q4 Adding another square will produce a similar result to adding the first square: The new, larger rectangle will have the same shape as the smaller starting rectangle.

A GOLDEN SPIRAL

13. The **Arc on Circle** command is available because points *C* and *E* are equally distant from *D*, defining an implicit circle centered at *D*.

14. It's important to hide the labels so that the tool students will make in step 15 does not clutter the sketch with labels every time it is used.

15. The tool will work correctly only in one direction, so students will have to take care what point they click first. Fortunately, they can always use **Edit | Undo**.

Q5 The new, larger rectangle also has its width and height in the golden ratio.

Q6 Answers will vary, but it's best if students make at least four or five new rectangles.

Q7 Each new rectangle is golden, with its sides in the golden ratio.

Q8 Even if the original rectangle is not golden, successive rectangles become closer and closer to golden in shape.

EXPLORE MORE

Q9 As the rectangles get larger, the ratio becomes closer and closer to the golden ratio.

Q10 The sizes of the first 10 squares are 1, 1, 2, 3, 5, 8, 13, 21, 35, and 56. These are the Fibonacci numbers. (The ratio between successive Fibonacci numbers gets closer and closer to the golden ratio.)

WHOLE-CLASS PRESENTATION

Use the **Golden Rectangle Present.gsp** sketch to demonstrate golden rectangles and golden spirals.

Use page 1 of the sketch to show the steps of the construction through step 12. After clicking the buttons, drag point C to make the ratios equal. On page 2 you can use a tool to build the first golden spiral, covering steps 13 through 16. Page 3 is for the Explore More construction. Page 4 shows the construction used to create the custom tool.

The Golden Rectangle and Ratio

 elongated square

The ratio of the width of a golden rectangle to its height is called the *golden ratio*.

You can describe the shape of a rectangle using the ratio of its width to its height. The rectangle on the left has a ratio of about 5:1, and the square on the right has a ratio of 1:1. The shape in the middle is often considered to be more attractive and has been called the *golden rectangle*. Paintings, photos, books, and magazines are often made with proportions similar to the golden rectangle.

If you add a square to the long side of a golden rectangle, the result is still a golden rectangle, oriented vertically rather than horizontally. The new, larger rectangle still has its sides in the same ratio as the original. You'll use this property to construct a golden rectangle, determine its ratio of width to height, and explore its characteristics.

SKETCH AND INVESTIGATE

You'll begin by constructing an adjustable rectangle.

1. In a new sketch, use the **Segment** tool to draw segment *AB*.

With the **Arrow** tool, select the segment and both points. Then choose **Construct | Perpendicular Lines.**

2. Construct two lines perpendicular to segment *AB*, one through point *A* and the other through point *B*.

3. Construct point *C* on the line through *B*.

4. Construct a line parallel to segment *AB* through point *C*. Construct intersection *D*.

5. Hide the lines and construct segments to connect the four points.

Select the points in order and choose **Construct | Quadrilateral Interior.**

6. Construct the quadrilateral interior.

Choose **Number | Calculate** to show the Calculator. Click the measurements in the sketch to enter them into the Calculator.

7. Measure the lengths of segments *AB* and *AD* by selecting the segments and choosing **Measure | Length.** Calculate the ratio of the width to the height.

m $\overline{AB}$ = 4.48 cm
m $\overline{AD}$ = 1.66 cm
$\dfrac{m\ \overline{AB}}{m\ \overline{AD}}$ = 2.70

Q1 What point must you drag to adjust the rectangle's shape? Drag that point to make the shape more attractive to you. What's the ratio now?

Now you'll add a square above the rectangle.

Use the Transform menu to mark a center and rotate a point.

8. Mark point D as the center of rotation, and rotate point C by 90° about D. Label the new point E.

9. Mark point E as the center of rotation, and rotate point D by 90°. Label the new point F.

10. You now have the four vertices of the added square. Construct the sides and interior of the square.

To measure the distance from A to E, select the two points and choose **Measure | Distance**.

11. The original rectangle and the new square together make a larger rectangle. Measure the longest side of this new rectangle. Then calculate the ratio of the longer and shorter sides of the large rectangle.

Q2 How does this ratio compare to the ratio from the original rectangle?

12. Adjust point C until the two ratios are as close to equal as you can make them.

The golden ratio is often represented by the Greek letter ϕ (phi).

Q3 What are the ratios now? This is the value of the golden ratio.

Q4 If you add another square on side AE to make a still larger rectangle, what do you think will be the ratio of the sides of this rectangle?

A GOLDEN SPIRAL

By constructing an arc inside square $CDFE$ and then adding more squares with arcs, you can construct a golden spiral.

13. Select in order points D, C, and E. Choose **Construct | Arc on Circle**.

14. Hide the labels of points C, D, F, and E.

Press and hold the **Custom** tool icon to show the Custom Tools menu.

15. Create a tool to make it easy to repeat the process: Select points C, D, F, and E, the segments connecting them, the square interior, and the arc. Then choose **Create New Tool** from the Custom Tools menu, and name the new tool **Square With Arc**.

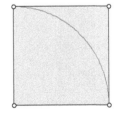

To use the new tool, choose it from the Custom Tools menu.

16. Use the new tool on points A and E to add a new square.

Q5 Measure the length and height of the new rectangle, and calculate their ratio. What result do you get?

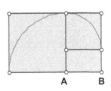

To make more squares, you'll have to make the original shapes smaller.

Make sure each new arc connects to the previous arc.

17. Move *A* and *B* closer together, and use the new tool several times to add more squares to the existing rectangles.

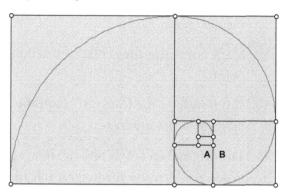

Q6 How many rectangles did you make?

Q7 What do you think is the ratio of the sides of the largest rectangle you made? Measure the distances and calculate the ratio to confirm your conjecture.

Q8 If your original rectangle was not a golden rectangle, what effect do you think there would be on the ratio of the sides of the largest rectangle? Drag point *C* to find out. What do you observe about the shape of the largest rectangle as you change the shape of the smallest one?

EXPLORE MORE

Q9 Start a new spiral by using the **Square With Arc** tool twice at the same size before you start adding larger squares. As the rectangles get larger, what happens to the ratio of the sides?

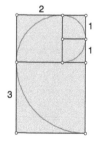

Q10 If the first square has sides of length 1, so does the second. How long are the sides of the third square? How long are the sides of the fourth square, and the fifth? Write down the sizes of the first 10 squares. Have you ever seen these numbers before?

Exploring Algebra 1 with The Geometer's Sketchpad
© 2012 Key Curriculum Press

Fractals and Ratios

The image beside the introductory paragraph is Barnsley's Fern. You can research it on the web.

RATIO OF SIMILARITY

The description of the term *similarity* here is rather vague. A more precise definition would mention the requirement that angles must be equal and corresponding distances must be in proportion. Consider addressing the meaning of this term in a class discussion.

Q1 Using fraction notation here would help to reinforce the concept of the ratio, but all of the ratio measurements will be in decimals.

    a. $9/10 = 0.9$ b. $81/100 = 0.81$

    c. $729/1000 = 0.729$ d. $81/100 = 0.81$

Q2 When you drag k, the positions of the posts change, but their heights and ratios remain constant.

ITERATION

Q3 Each iteration moves only one-tenth of the remaining distance to the point, so no finite number of iterations will bring it to the point. (Because of the limited resolution of computer screens, students may think that the fractal does reach point C after a large number of iterations. Try to elicit from students convincing arguments that the fractal can never reach point C.) This is a good illustration of the concept of limitless complexity. Because of this limitless complexity, you cannot construct an entire fractal. You can only define rules for the drawing and carry them out until the detail gets too small to see.

If you could zoom in on point C, you would see the same image. Using the fence post analogy again, this is like driving another mile down the road. The row of posts will still look the same. Point C is on the horizon, and you can never reach the horizon.

Q4 The rectangular spiral is a rearrangement of the parts of the fence post fractal. The lengths of the new segments are the same as those in the fence post fractal, but each is perpendicular to the previous segment rather than parallel to it.

 ACTIVITY NOTES

SELF-SIMILARITY

Q5 When *DE* is matched to one of the line segments of the black fractal, the red fractal matches all of the higher levels. This demonstrates that small parts of the black fractal are similar to the whole.

Q6 1.0 (level 0) $9/10 = 0.9$ (level 1)

$81/100 = 0.81$ (level 2)

$729/1000 = 0.729$ (level 3)

$6561/10000 = 0.6561$ (level 4)

Q7 1.0 (level 0) $7/10 = 0.7$ (level 1)

$49/100 = 0.49$ (level 2)

$343/1000 = 0.343$ (level 3)

$2401/10000 = 0.2401$ (level 4)

WHOLE-CLASS PRESENTATION

Open **Fractals and Ratios Present.gsp.** Press the button to show the second fence post, and then show the ratio of their heights $(k':k)$. Use the buttons to show more fence posts. Have students predict the ratios (from Q1 of the student activity), and then use the buttons to show them.

Similarly, use the provided buttons to show the iterated fence posts on page 2.

Fractals and Ratios

A *fractal* is a geometric figure with limitless complexity. Whenever you look at some small detail, you find an even smaller detail. The fractals you will investigate here are also *self-similar*. This means that small parts of the figure are actually scale replicas of the entire figure.

RATIO OF SIMILARITY

First, you will look at the similarity of line segments. Two objects are *similar* if they have the same shape but not necessarily the same size. All line segments are similar because they all have the same shape. The *ratio of similarity* between two line segments is the ratio of their lengths.

1. Open **Fractals and Ratios.gsp.** The first page contains only line segment k and point C. Think of this as the first post in a fence that continues along a highway toward point C. If you take a photograph of the fence, the second post will appear smaller in the distance.

To mark the center, select the point and choose **Transform | Mark Center.**

2. Mark point C as the center for dilation.

3. Dilate segment k by selecting it and choosing **Transform | Dilate.** In the dialog box that appears, define the scale factor as the ratio 9/10. A new, smaller segment appears, representing the next fence post.

To show the labels, select the segments and choose **Display | Show Labels.**

4. Leaving the new line segment selected, dilate it by the same ratio. Continue dilating until there are at least four line segments on the screen. Show the labels of all the segments.

5. Drag k left and right and observe the effect on the dilated images.

The length of line segment k' is 9/10 the length of line segment k. This is the ratio of similarity of $k' : k$. You can confirm this ratio by measuring it.

To measure the ratio, select in order k' and k. Then choose **Measure | Ratio.**

6. Measure the ratio of k to k.

Q1 What are the following ratios of similarity? Calculate them on paper. Then check your answers by choosing **Measure | Ratio.**

    a. $k'':k'$ b. $k'':k$

    c. $k''':k$ d. $k''':k'$

Q2 Drag k again. Do the measurements change?

ITERATION

Next you will use iterated constructions to draw some fractals. In the previous construction, you repeated a simple dilation several times. Each repetition is an *iteration*. Now you will have the computer do the iterations for you.

7. Go to the "Fence Post 2" page. In this sketch, the endpoints of the line segment have been dilated once. You will use the parameter *depth* to control the number of iterations.

8. Select in order point A, point B, and parameter *depth*. Hold down the Shift key and choose **Transform | Iterate to Depth.**

9. A dialog box appears asking how to map the two points. Answer by clicking the correct points in the sketch. Map point A to D, and map B to E. From the Structure pop-up menu, choose **Non-Point Images Only.** Click **Iterate.**

10. To increase the depth of the iteration, select the parameter *depth* and press the **+** key.

> You can use Sketchpad's **Dilation Arrow** tool to zoom in on point C. Double-click C to mark it as the center, and then drag the fence posts away from C.

Q3 As you increase the depth, each new level of the fractal is closer to point C. How many iterations does it take for the fractal to reach point C? What would you see if you "zoomed in" and got a closer look at the fence closer to point C?

The figure below shows the first few levels of a different fractal. It begins as a single line segment. This line segment is rotated 90° about one endpoint, and then dilated by a ratio of 9/10. That brings it to level 1. To reach level 2, the same thing is done to the new segment. Each new iteration adds one line segment.

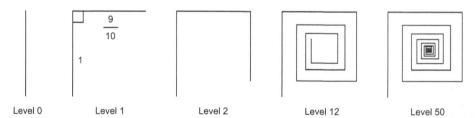

    Level 0   Level 1   Level 2   Level 12   Level 50

Q4 How is this fractal different from the fence post fractal? How is it the same?

11. Open the "Rectangular Spiral" page. You will see the original segment and the endpoints of the second one.

12. Select in order point *A*, point *B*, and parameter *depth*. Hold down the Shift key and choose **Transform | Iterate to Depth.** This time map *A* to *B* and *B* to *C*. As before, choose **Non-Point Images Only.** Click **Iterate.** Increase the depth to at least 20.

SELF-SIMILARITY

13. Press the *Show Similar Fractal* button.

Q5 Press the red *Level* buttons to align segment *DE* with different levels of the fractal that you constructed. Describe what you see. How does this relate to self-similarity?

Q6 Measure the ratio *DE/AB.* Use the red buttons again and record the ratio of similarity for levels 0 through 4 of your fractal.

Q7 The construction is based on the ratio 9/10. Change it to 7/10. Now what is the ratio of similarity for levels 0 through 4? Use the red buttons to check your answers.

Length of the Koch Curve

 ACTIVITY NOTES

This activity reinforces and reviews various operations involving fractions and exponents while giving students a chance to construct an interesting and beautiful fractal at the same time. You can use the activity during a review of fractions and exponents, as part of a unit on fractals, or as an engaging enrichment activity.

WHAT IS THE KOCH CURVE?

Q1 Each level 1 segment is 1/3 unit long.

Q2 The entire level 1 curve is 4/3 units long.

Q3 Each level 2 segment is 1/9 unit long. There are 16 such segments, so the entire curve is 16/9 units long.

Q4 At this point students are making conjectures. Try to use the conjectures to get students to discuss and explain their reasoning, without emphasizing the "right" answers. (In fact, it does get longer at each level, and there is no limit to its length. There are limits to its range: It would never outgrow the page.)

Q5

Level	# Segments	Segment Length	Total Length
0	1	1	1
1	4	1/3	4/3
2	16	1/9	16/9
3	64	1/27	64/27
4	256	1/81	256/81

Q6 The curve length at level 10 is $(4/3)^{10}$, which is about 18. The general formula for the length at level n is $(4/3)^n$.

Q7 There is no limit to the length of this fractal, which is interesting because the idea of a curve of unbounded length within a bounded region is counterintuitive. (Not all fractals have unbounded length. See the fence post and rectangular spiral fractals in the activity Fractals and Ratios.)

EXPLORE MORE

Q8 The length of the meander also grows at a geometric progression, but with ratio 3/2. At level n, the length of the fractal is $(3/2)^n$. As with the Koch curve, its length is without limit, but its range is limited.

WHOLE-CLASS PRESENTATION

Use the presentation sketch **Koch Curve Present.gsp** to demonstrate the construction and characteristics of the Koch curve. Then develop the table for Q5 as a class activity.

RELATED PROJECTS

P1 Research other fractals. Include in your report other ways of describing fractals and pictures of interesting fractals that you find.

P2 Research the history of fractals. When were fractals first described, and by whom? How are they used in movies and other forms of computer-generated graphics?

P3 Find a picture of the Sierpiński triangle, and try to figure out how to construct it using Sketchpad.

P4 Find other fractals that can be constructed with Sketchpad, construct them yourself, and present your constructions to your group or class.

P5 Research the term *fractal dimension,* and determine the fractal dimension of the Koch curve and of the meander.

Length of the Koch Curve

When a fraction of a shape is similar to the entire object, the shape is called a *fractal*.

If you magnify a small fraction of an ordinary curve such as a circle, the small fraction appears straighter than the curve as a whole. This isn't true of fractals. For example, the Koch curve is always similarly bumpy, no matter how much you magnify it. In this activity you will create a Koch curve and investigate its length.

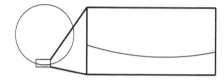

WHAT IS THE KOCH CURVE?

The Koch curve was first created by Swedish mathematician Helge von Koch (1870–1924).

The easiest way to describe a Koch curve is by using a *recursive* rule—a rule that is applied over and over again. Start with a segment (level 0) and divide it into thirds. Remove the middle third and replace it with two new segments, each equal in length to the removed segment (level 1). Apply this rule again to each new segment to see the next level of the Koch curve.

| Level 0 | Level 1 | Level 2 | Level 3 |

1. Open **Koch Curve.gsp.** This sketch already has the level 0 curve and the points you need to make the level 1 curve.

To hide the long segment, select it and choose **Display | Hide Segment.**

2. Hide the level 0 curve (the long segment).

3. Construct the four segments of the level 1 curve by using the **Segment** tool.

Q1 If the original level 0 segment is 1 unit long, how long is each level 1 segment?

Q2 How long is the entire level 1 curve?

Creating many levels by hand would be time-consuming and error-prone. Instead, you will use Sketchpad's iteration feature to produce the levels automatically.

This page gives you a fresh level 0 starting place.

4. Go to the "By Iteration" page of the document.

5. Select in order point *A*, point *B*, and parameter *depth*. Hold down the Shift key and choose **Transform | Iterate to Depth.**

6. In the dialog box that appears, specify the endpoints of the first segment on which to iterate the construction. Start with the segment on the left, by clicking in the sketch on points *A* and *C*. The points appear in the First Image column.

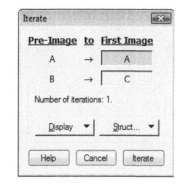

7. To apply the same construction to points *C* and *D*, you must add a new map. Click the Structure button in the dialog box. In the pop-up menu that appears, choose **Add New Map.** Then click points *C* and *D* in the sketch.

8. Continue adding maps and mapping the points until you have done the construction on all four segments.

Pre-Image	To	Map #4	Map #3	Map #2	Map #1
A	→	*E*	*D*	*C*	*A*
B	→	*B*	*E*	*D*	*C*

To change the parameter, select it and press the + or − key. Don't go past 6 or your computer will slow down.

9. From the Display pop-up menu, choose **Final Iteration Only.** Then click **Iterate.**

10. Press the *Hide Level 0* button to hide the original segment and construction points. Change the depth parameter to see a higher level of the curve.

Q3 How long is each segment of the level 2 curve? How long is this entire curve?

Q4 Does this curve get longer at each level? If you keep applying the rule, how long will it eventually become? Will it run off the page?

Next you will compare the levels of the Koch curve.

Q5 On paper, make a table comparing levels of the curve from zero through four. How many line segments are there on each level? How long is each of them? What is the total length of the curve? Include this information in your table.

You can answer this more easily using exponents.

Q6 What is the total length of this curve at level 10?

Q7 Is there any limit to the length of this fractal?

EXPLORE MORE

Q8 On the "Meander" page is the setup for the meandering fractal below. This fractal also starts with a segment that is 1 unit in length. For each level after level 0, you replace each segment with three segments that are half as long. This iteration requires three mappings. Do the construction and answer the same questions (Q1–Q7) about its length.

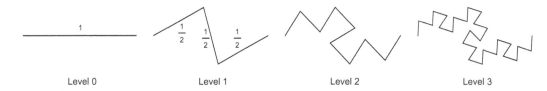

| Level 0 | Level 1 | Level 2 | Level 3 |

The Chaos Game

Students should always be encouraged to explain their answers, but many of the things they observe in this activity will be difficult for them to fully grasp. You should expect them to write (or draw) a clear description of a pattern even if they are unable to explain why it appears.

THE SETUP

Q1 P' is in line with P and one of the triangle vertices.

Q2 Dragging point r changes the ratio r. When $r = 0$, P' is at P. When $r = 1$, P' is at one of the triangle vertices. The distance PP' is the distance from P to the vertex scaled by ratio r.

Q3 When you move point k, P and P' align with a different vertex of the triangle. The path of k is a line segment. The line segment is divided into thirds. Each third corresponds to one vertex of the triangle.

THE GAME

2. Demonstrate this step if possible.

3. This step calls for a depth of 1000, which might be a bit conservative. Some computers are fast enough to use much greater depth. (You can use **Edit | Advanced Preferences** to change the upper limit.) See what your computers can tolerate, and advise your students.

Q4 The pattern is similar to that of a Sierpiński triangle, which the students may have seen before.

In general, the position of point P appears to affect the first few points of the orbit, but there is very little difference in the overall pattern. In fact, changing P changes all the points, but the change is too small to notice in any but the first few iterations.

Q5 When $r = 0$, the pattern completely disappears. This is because P is mapped to itself on every iteration. The entire orbit is occupying only that one point.

When $r = 1$, the pattern disappears again, but for different reasons. Point P is mapped to a vertex. Throughout the orbit, it is jumping between the three vertices, but never anywhere else.

Q6 As *r* decreases, the triangle patterns overlap and the open regions are covered. When *r* reaches 0.33 (1/3), all of the open regions are covered and the triangle pattern is no longer visible.

Q7 When $r < 0$ or when $r > 1$, the orbit leaves the triangle. At values slightly higher than 1, there is a pattern with rotational symmetry.

When $r > 2$, the pattern changes. Point *P* essentially jumps over the vertices and ends up farther away and on the other side.

Q8 When $r = 1.99$, the orbit tends to stay near the triangle, and it forms a pattern with something close to 180° rotation symmetry.

When $r = 2.00$, the rotation symmetry is still there, but the orbit does not seem to be either attracted or repelled by the triangle. Also, close inspection will show that the points fall in a triangular grid, like isometric dot paper.

When $r = 2.01$, the pattern diverges in two different directions.

EXPLORE MORE

Q9 All of the observations from the previous sections hold true for any triangle.

Q10 The instructions are the same. Look for the Koch snowflake that appears on the Hexagon page.

WHOLE CLASS PRESENTATION

To present this activity to the entire class, follow the Presenter Notes and use the sketch **Chaos Game Present.gsp.**

Exploring Algebra 1 with The Geometer's Sketchpad
© 2012 Key Curriculum Press

The Chaos Game

In this presentation students observe a process in which a point is repeatedly dilated toward a randomly chosen point by a particular ratio. They see how such a random process can sometimes generate a regular pattern, depending on the ratio used.

1. Open **Chaos Game Present.gsp.**

Have several students answer each question in their own words.

Q1 Drag point P across the screen. Ask, "How is point P' related to P?"

Q2 Drag point r along its segment. Ask, "How does the value of r affect the relationship between P' and P?"

Q3 Drag point k. Ask "What effect does k have on the sketch?"

Q4 Ask, "What do you think would happen if we repeated this process starting with point P'?" Be sure to get several different predictions.

2. Press the *Show Iterated Image (constant k)* button to show the next image.

Q5 Ask, "What would happen if we did this over and over again?"

Use the + sign to increase depth, and the – sign to decrease it.

3. To test students' predictions, select parameter *depth* and press the **+** sign on the keyboard several times.

4. Hide the iterated image by pressing the *Hide Iterated Image (constant k)* button. Set *depth* back to 1 by pressing the **–** sign or by double-clicking the parameter and changing its value.

Q6 Ask, "What would happen if we did the same thing, but chose a random vertex at each step?" Get several different predictions concerning the pattern.

5. Press the *Show Iterated Image (random k)* button to show the next image. Increase the depth slowly, so students can observe how the next point goes halfway toward a randomly chosen vertex at each step.

Q7 Ask, "Can you see a pattern yet?" Students may or may not be able to make a prediction yet.

To stop the animation, press the Animate Depth button again.

6. Press the *Animate Depth* button, and let the animation run for a while. Stop when a pattern begins to emerge.

Q8 Have students report the pattern that they see. The greater the depth, the clearer the pattern will become. (Be careful increasing the depth; if you increase it too much, your computer will slow down.)

Set depth to a value at which your computer does not slow down too much.

Q9 Ask, "What do you think will happen if r is greater than 0.5? What if r is less than 0.5?" Drag r to investigate both situations. Ask students to explain the patterns they see based on the value of the ratio.

Use the remaining pages of the sketch to investigate patterns produced by similar iterations involving different numbers of vertices.

The Chaos Game

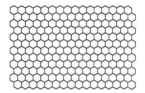

Patterns happen, often with no effort to make a particular pattern. If you mow a lawn, you will likely end up with a pattern in the grass. Bees have no notion of a hexagon but create a hexagonal tiling when they nestle their wax chambers as closely as possible. The Chaos Game is a study of some patterns that result from random choices.

THE SETUP

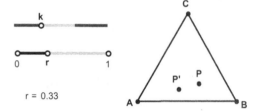

1. Open **Chaos Game.gsp.** Triangle ABC is an equilateral triangle. Point P is an independent point.

Q1 Drag point P across the screen. How is point P' related to P?

Q2 Point r controls ratio r. What changes occur when you drag point r? Where is P' when $r = 0$? Where is P' when $r = 1$? How is r related to the distance PP'?

Q3 Drag point k. What effect does that have on the sketch? Describe it in detail.

THE GAME

The idea of the Chaos Game is to start P moving and guess where it will go. First it goes to P'. From there, it will use the same ratio, r, and go toward another vertex. Which vertex? That's the random part. It could be any of the three.

2. Set the parameter *depth* to 1 and r to 0.50. Select in order points P and k and parameter *depth*. Hold down the Shift key and choose **Transform | Iterate to Depth.** A dialog box appears asking where the two points should be mapped. Answer by clicking the correct points. Map P to P', and map k to k. From the Structure pop-up menu, choose **To New Random Locations.** Click **Iterate.**

To animate the parameter *depth*, select it, choose **Display | Animate,** and use the Motion Controller to control the animation speed.

3. There is a new point on the screen. To see the next point, increase *depth* to two. Animate the *depth* parameter and observe the path as the point moves to new locations. This path is called the *orbit* of the point. Run the animation until *depth* is 1000 or more.

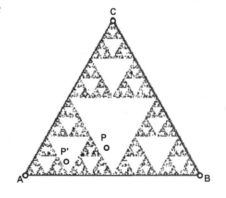

To change the parameter in one step, double-click the parameter and enter the value.

4. Select the orbit. Choose **Point Style | Dot.** This will make the points smaller so that the pattern is easier to see.

Q4 Describe the pattern made by the orbit when $r = 0.50$. Drag point P. What effect does that have on the pattern?

You can still change many of the conditions after you have plotted the orbit. Drag point B to change the size of the triangle, drag r to change the ratio, and place P anywhere you wish. Following is a list of conditions that you can create. In each case, describe what you see and explain what causes the image to appear the way it does.

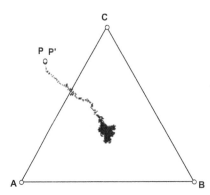

Q5 What does it look like when $r = 0$? What does it look like when $r = 1$? Explain what causes these orbit patterns.

To see a different random pattern, select the orbit and press the exclamation point (!) on your keyboard.

In order to make fine adjustments to r, select point r and use the left and right arrow keys on the keyboard.

Q6 Start with point P inside the triangle and $r = 0.50$. There are many triangle patterns in the orbit. Slowly decrease r. What is r when the triangle patterns disappear?

Q7 Start with point P inside the triangle. For what range of r does the orbit leave the triangle?

Q8 Drag point B so that the triangle is very small, only a few pixels across. Set r to 1.99, then 2.00, and then 2.01. Describe the orbit.

EXPLORE MORE

So far, you have been working with an equilateral triangle. See what happens when you use some other triangle.

5. Select point B. Choose **Edit | Split Point from Ray.**

6. Select point C. Choose **Edit | Split Intersection from Circles.**

Q9 Now all of the vertices are independent points. Go back through the previous settings and see if your observations apply to other triangles.

Q10 In the same document, there are other pages with a square, a regular pentagon, and a regular hexagon. The rest of the setup is the same. Perform steps 2–4 again on these pages and see what patterns you can find.

Exponents

MULTIPLICATION PATTERNS

Q1 Answers will vary. This question encourages students to observe closely and think about the patterns they see. The points represent various powers of x, with the vertical position corresponding to the power and the horizontal position corresponding to the value of x.

Q2 There are two positions of x at which all the points line up: $x = 1$ and $x = 0$. Repeatedly multiplying either 1 or 0 by itself continues to give the same result.

Q3 When $x > 1$, the points move increasingly rightward as you go down the screen, showing that the value of x^n increases more and more quickly for larger and larger values of n. Be sure students notice that the differences in the lengths of the bars are not constant, but increasing.

Q4 When $0 < x < 1$, the points move toward the left as they go down, approaching a straight line (vertical asymptote) at $x = 0$. This pattern makes sense because multiplying by a value less than one always gives a result that is closer to zero than the number you started with. Be sure students notice that the differences in the lengths of the bars are not constant, but decreasing.

Q5 When $x < 0$, the bars alternate between the left and right sides, because multiplying a number by a negative value always gives a result with a sign opposite to the sign of the original number.

MULTIPLYING AND DIVIDING

Q6 The $x^3 \cdot x^2$ bar is the same length as the x^5 bar no matter how you drag the value of x. This makes sense, because you've multiplied three x's by two more x's, so that there are now five x's multiplied together.

Q7 The bar for $x^4 \cdot x^3$ is the same length as the x^7 bar.

Q8 Conjecture: The bar for $x^a \cdot x^b$ is the same length as the x^{a+b} bar. Students will construct different problems to test this conjecture.

Q9 When you use the **a / b** tool to create a bar for x^7 / x^3, the resulting bar is the same length as the x^4 bar.

Q10 Conjecture: The bar for x^a / x^b is the same length as the x^{a-b} bar. Students will construct different problems to test this conjecture.

. . . .

RAISING TO A POWER

Q11 When you use the **(a)^b** tool to create a bar for $(x^4)^2$, the result matches x^8.

Q12 Conjecture: The bar for $(x^a)^b$ is the same length as an x^{ab} bar. Students will construct different problems to test this conjecture.

Q13 $x^a \cdot x^b = x^{a+b}$: Because the first factor (x^a) represents a values of x multiplied together, and the second (x^b) represents b values of x multiplied, the product has $a + b$ values all multiplied together.

$x^a / x^b = x^{a-b}$: When you divide, the first b factors of x in the numerator will cancel out with the factors in the denominator, leaving the $a - b$ factors of x as the result.

$(x^a)^b = x^{ab}$: The second exponent (b) means that there are b factors to multiply together, where each factor is x^a. The total number of factors of x is ab, so this is the exponent in the result.

EXPLORE MORE

Q14 If the bases are not the same, there is no rule you can use to simplify a problem like $x^a \cdot y^b$. Although there are some special cases that may suggest a more general result, testing with many values of x, y, a, and b will show that there is no general pattern.

Exponents

In this presentation students use a visual representation of powers of x to understand and explain the exponent rules that apply to the expressions $x^a \cdot x^b$, x^a / x^b, and $(x^a)^b$.

1. Open **Exponents.gsp.** Drag the red point (x) left and right.

Q1 Ask, "What do you observe? What do you think the points represent?"

Q2 Continue dragging slowly, and ask, "How many positions are there that make all the points line up? What values of x do you think these positions indicate?" Have students explain why the points should line up at these particular values.

2. Click the *Show Constants* button and the *Show Bars* button, and drag x again.

Are the differences in length constant from one bar to the next?

Q3 Ask students to describe and explain the patterns they observe when $x > 1$, when $0 < x < 1$, and when $x < 0$.

3. Remind students that the shortcut for an expression like $x \cdot x \cdot x \cdot x \cdot x \cdot x \cdot x$ is x^7. Then go to the "Multiplication" page of the sketch. Drag point x left and right to make sure students observe the same patterns as they did on page 1.

In the next several steps, you'll demonstrate what happens when you multiply or divide two of the values represented by the bars.

4. Choose the **a*b** custom tool from the Custom Tools menu. Explain that this tool allows you to multiply any of the exponent bars.

5. Use the **a*b** tool to multiply x^3 by x^2, by clicking 5 objects in order: the first unused point below all the bars, the point at the tip of the x^3 bar, the x^3 caption, the point at the tip of the x^2 bar, and the x^2 caption.

Q6 Ask, "Is the $x^3 \cdot x^2$ bar the same length as any of the existing bars?" Drag x back and forth to give students a chance to observe. [They should answer x^5.]

6. To confirm this answer, choose the **Indicator** custom tool and click it on the tip of the $x^3 \cdot x^2$ bar. Then use the **Arrow** tool to drag x left and right.

7. Create a bar for $x^4 \cdot x^3$, and ask students what existing bar it matches. [x^7]

Q7 Have students write out both x^4 and x^3 using repeated multiplication, and ask them how this way of writing it explains the way the bars behave.

Q8 Ask students to make a conjecture concerning the result of $x^a \cdot x^b$, and to explain why the conjecture makes sense.

Use the "Division" and "Power" pages to investigate x^a / x^b and $(x^a)^b$.

Exponents

In this activity you'll use a Sketchpad model to perform operations involving exponents.

MULTIPLICATION PATTERNS

1. Open **Exponents.gsp.**

2. Drag the red point (x) left and right, and observe the motion of the other points.

   Q1 What do you think the points represent?

   Q2 How many positions can you find that make all the points line up? What values of x do you think these positions indicate? Explain why the points should line up at these particular values.

> Click the *Show Constants* and *Show Bars* buttons to show these objects.

> Are the differences in length constant from one bar to the next?

3. Show the constants and the bars, and then drag point x again to observe the behavior of the bars and to check your results from Q1.

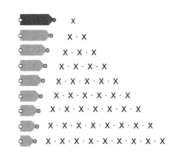

   Q3 What pattern do the points make when $x > 1$? Why does this pattern make sense?

   Q4 What pattern appears when $0 < x < 1$? Explain this pattern.

   Q5 What pattern appears when $x < 0$? Explain this pattern.

EXPONENTS

With addition, when you have a problem like $x + x + x + x + x + x + x$, you can write the problem more easily using multiplication:

$$x + x + x + x + x + x + x = 7x$$

Similarly, when you have a multiplication problem like $x \cdot x \cdot x \cdot x \cdot x \cdot x \cdot x$, you can write the problem more easily using an exponent:

$$x \cdot x \cdot x \cdot x \cdot x \cdot x \cdot x = x^7$$

4. Go to the "Multiplication" page. This page uses exponents to write the multiplication problems more simply. Drag point x left and right to make sure you observe the same patterns as you did on page 1.

MULTIPLYING AND DIVIDING

In the next several steps, you'll investigate what happens when you multiply or divide two of the values represented by the bars.

5. Choose the **a*b** custom tool from the Custom Tools menu. This tool will allow you to multiply any of the exponent bars.

6. Use the **a*b** tool to multiply x^3 by x^2, by clicking 5 objects in order: the first unused point below all the bars, the point at the tip of the x^3 bar, the x^3 label, the point at the tip of the x^2 bar, and the x^2 label.

Q6 Is the $x^3 \cdot x^2$ bar the same length as any of the existing bars? Test your answer. Click the **Indicator** custom tool on the tip of the $x^3 \cdot x^2$ bar. Then use the **Arrow** tool to drag x left and right. Describe your observations.

Q7 Create a bar for $x^4 \cdot x^3$, and determine what existing bar it matches.

When you test, be sure to drag x left and right to try different values.

Q8 Make a conjecture concerning the result of $x^a \cdot x^b$, and test your conjecture by constructing another multiplication problem.

7. Go to the "Division" page. Use this page to explore division problems.

To use this tool, click it on 5 objects in order, just as you did with the **a*b** tool.

Q9 Use the **a / b** tool to create a bar for $\frac{x^7}{x^3}$. What is the result?

Q10 Make a conjecture concerning the result of $\frac{x^a}{x^b}$, and test your conjecture by constructing another division problem.

RAISING TO A POWER

8. Go to the "Power" page. You will use this page to explore problems like $(x^4)^2$.

To use this tool, click the starting position, the point and label for x^4, and the point and label for the constant 2.

Q11 Use the **(a)^b** tool to create a bar for $(x^4)^2$. What is the result?

Q12 Make a conjecture concerning the result of $(x^a)^b$. Test your conjecture by constructing another similar problem.

Q13 Summarize your conjectures from Q8, Q10, and Q12. Include an example for each conjecture. Use the definition of exponents (as repeated multiplication) to explain why each of your conjectures makes sense.

EXPLORE MORE

Q14 What if the bases are not the same? Is there a rule you can use for problems like $x^a \cdot y^b$? Go to the "Explore" page and experiment. Describe your conclusions.

Zero and Negative Exponents

POSITIVE EXPONENTS

2. The tools will work for any nonzero setting of *a*, but the activity creates an exponential sequence, so the heights of the rectangles can get out of hand if *a* is much more than 1.5.

Q1 $a^2 \cdot a = a^3$

Q2 Multiplying by *a* raises the power by one: $a^n \cdot a = a^{n+1}$. This is true because $a^n = a \cdot a \cdot a \cdots a$, where *a* appears as a factor *n* times. If you multiply this by one more *a*, there will be $n + 1$ factors.

Q3 As the exponents increase, the heights of the bars do not increase by the same amount each time. Provided $a > 1$, they increase by more and more each time. This is more obvious if you change the scale by dragging one of the tick numbers on the axis. Students may use specific examples to explain: "Multiplying 1 by 1.37 adds only 0.37 to the bar height, but multiplying 10 by 1.37 results in 13.7, adding 3.7 to the original height."

ZERO EXPONENT

Q4 $a^4 \div a = a^3$, and generally, $a^n \div a = a^{n-1}$. As in Q2, $a^n = a \cdot a \cdot a \cdots a$, where *a* appears as a factor *n* times. If you divide this by *a*, you cancel the last factor, leaving $(n - 1)$ factors.

Q5 $a^0 = 1$. This follows from the fact that $a^1 \div a = 1$.

NEGATIVE EXPONENTS

Q6 $a^0 \div a = a^{-1}$, using the rule from Q3.

Q7 Since $1 = a^0$, dividing by *a* three times is equivalent to dropping the exponent three times to a^{-3}. Therefore, $1/a^3 = a^{-3}$.

EXPLORE DIFFERENT BASES

In this extension students can use the same sketch to see the effects of changing the base.

Q8 If $a > 1$, higher exponents always correspond to larger rectangles, hence, larger numbers. As you keep multiplying, there is no limit to how high the bars will go. Similarly, if you divide repeatedly, there is no limit to how short the bars will become.

Q9 If $0 < a < 1$, higher exponents correspond to smaller numbers. This follows from the fact that multiplying any positive number by a number greater than one increases it, while multiplying it by a number between one and zero makes it smaller. In either case, the result is always positive.

Q10 As a changes, any bar with the value a^0 remains constant. This is because $a^0 = 1$.

Q11 When a is less than zero, the bars alternate between positive and negative. The numbers with odd exponents are negative. Those with even exponents are positive.

In each step of the activity, students formed the next number by either multiplying or dividing by a negative number, a. Multiplying or dividing by a negative changes the sign, thus creating the alternating pattern.

Q12 Problems and solutions to the Simplify game vary.

(Both the activity document and the presentation document have another custom tool, **Measure Bar,** which was not used in the activity. Choose the tool and click one of the bars. It will display the number represented by the bar.)

Exploring Algebra 1 with The Geometer's Sketchpad
© 2012 Key Curriculum Press

In this presentation students observe the visual pattern formed when an exponent increases as a number is repeatedly raised to higher powers, and observe the related pattern as the exponent is reduced first to zero and then to negative values.

> Explain that the label on the bar is a^1, which is the same thing as a.

1. Open **Zero Exponents Present.gsp.** Drag the marker so students can see how it changes the value of a. Return the marker to its original position.

2. To multiply the value represented by the first bar by a, press and hold the **Custom** tool icon to display the Custom Tools menu. Choose the **Multiply By a** tool. This tool works by itself, so there is no need to click anything.

Q1 Ask students what the new bar represents. Drag the marker to change the value of a to 2, so that students can see the new bar has the value 4. Return the marker to its original position before continuing.

Q2 Ask, "What will be the result if we multiply a^2 by a?"

3. Choose **Multiply By a** again. Use the tool several times.

Q3 Ask, "As the exponents increase, do the heights of the bars increase by the same amount each time? How can you tell?"

4. Choose the custom tool **Divide By a.**

Q4 Ask students to explain why the new bar is the height that it is.

5. Use **Divide By a** several more times, until the progression runs down to a^0.

Q5 Ask, "What do you think is the value of a^0?" Drag the marker so that students can see that the value of a has no effect on this bar.

Q6 Ask, "What will happen if we divide by a again?"

6. Choose **Divide By a.**

Q7 Discuss what the resulting bar represents. Choose **Divide By a** twice more during the discussion. Try to get students to propose formulations like these:

$$a^{-1} = 1 \div a = \frac{1}{a} \quad \text{and} \quad a^{-3} = 1 \div a \div a \div a = 1 \cdot \frac{1}{a} \cdot \frac{1}{a} \cdot \frac{1}{a} = \frac{1}{a^3}$$

Use the other numbered pages to create different patterns that involve both positive and negative exponents.

Use the "Simplify" page to give students practice in manipulating expressions to eliminate negative exponents.

Zero and Negative Exponents

By now you should be comfortable doing calculations with exponents that are positive integers. From here, certain questions naturally arise. What if the exponent is zero? What if it is negative? What if it is not an integer? This activity explores the concept of zero and negative exponents. Non-integer exponents will have to wait.

POSITIVE EXPONENTS

1. Open **Zero Exponents.gsp.**

The bar represents the number a. You can drag the marker to change its value. The label on the bar is a^1, which is the same thing as a.

2. Start with a between 1 and 1.5. You can change it later. Now multiply a^1 by a. Press and hold the **Custom** tool icon to display the Custom Tools menu. Choose the **Multiply By a** tool. This tool works by itself, so there is no need to click anything.

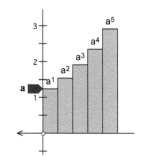

Another vertical bar appears representing a^2. You get this result because $a^1 \cdot a = a^2$.

Q1 What is the result if you multiply a^2 by a?

3. Choose **Multiply By a** again. Use the tool several times and study the result.

Q2 Consider the number a^n, where n is a positive integer. What happens when you multiply the number by a? State a general rule. Explain why this is true.

Q3 As the exponents increase, do the heights of the bars increase by the same amount each time? How can you tell? Explain your observations.

ZERO EXPONENT

4. Go to page 2. This is the same sketch. The progression of bars goes up to a^4.

5. Choose the custom tool **Divide By a.** It does just what the name says.

Q4 What is $a^4 \div a$? What happens when you divide a^n by a? Explain why this is true.

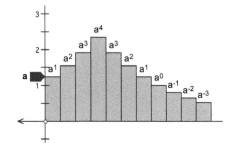

6. Use the **Divide By a** tool two more times, so that the progression runs down to a^1. Dividing by a once more should give you a^0. Try it.

Q5 What is the value of a^0? Drag the marker to test different values of a.

NEGATIVE EXPONENTS

Q6 At this point, a^0 should be the last number in the progression you are building. Using your answer to Q3, what will be the result when you divide by a again? Choose **Divide By a** and check your answer.

Q7 Starting with the number 1, if you divide by a three times, that is the same as dividing by a^3.

$$1 \div a \div a \div a = 1 \cdot \frac{1}{a} \cdot \frac{1}{a} \cdot \frac{1}{a} = \frac{1}{a^3}$$

How can you write this same number with a negative exponent? Use the sketch to check your answer.

EXPLORE DIFFERENT BASES

The point of this activity was to investigate zero and negative exponents, but you may have noticed some interesting changes that occur when you change the base, a. Drag the marker to change the value of a, and answer the following questions.

Q8 Start with $a > 1$. If you keep multiplying the bar lengths by a, is there a limit to how high the bars will go? If you keep dividing them by a, is there a limit to how short the bars will become?

Q9 Pull the marker downward so that $0 < a < 1$. What change do you see in the pattern formed by the bars? Explain why this is.

Q10 As you change a, which bars do not change at all? Why?

Q11 When a is negative, you will see an entirely different pattern. Describe the pattern, and explain why it looks this way.

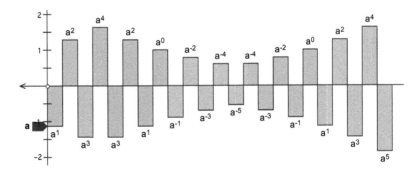

Q12 Play the game on the "Simplify" page at least four times. Each time, write down the original problem and the solution. Try writing the solution down before you actually drag the variables.

3

Algebraic Expressions

Order of Operations

Students consider the possibility that fictitious galactic cultures have different rules for evaluating expressions. The resulting discrepancies illustrate the importance of agreeing on and following standard rules for order of operations.

Equivalent Expressions

Students compare algebraic expressions and explore the equivalence of those expressions using "algebars," a model that shows the changing values of variables and expressions by the varying lengths of bars.

Equivalent Expressions: The Border Problem

Students invent a variety of equivalent expressions for a real-world area problem.

The Distributive Property: A Painting Dilemma

Students calculate an area in two different ways and verify that both ways give the same result. They generalize the calculations using variables instead of numbers, resulting in the distributive property for multiplication over addition. Students can explore the problem with measurements in meters or in feet.

The Distributive Property

Students use an interactive model to explain why the distributive property is true and to test whether the distributive property still holds if you interchange the role of addition and multiplication.

Algebra Tiles

Students model expressions as algebra tiles. They then use the areas of the tiles to see the equivalence of expressions in factored and expanded form.

The Product of Two Binomials

Students use Sketchpad algebra tiles to model the expanding of binomials. Students then use the results to demonstrate the principles of equivalent expressions.

Squaring Binomials

Students use algebra tiles to model the identities $(x + y)^2 = x^2 + 2xy + y^2$ and $(x - y)^2 = x^2 - 2xy + y^2$. They connect the geometric models and the terms that make up the algebraic expressions.

Squares and Square Roots

As students manipulate squares modeled on Sketchpad dot paper, they explore properties of squares and square roots, discover interesting number patterns, and explore the concept of rational versus irrational numbers.

DIFFERENT STROKES FOR DIFFERENT FOLKS

Q1 The systems differ. The Terran result is 7, but both Vulcans and Klingons get a result of 8.

Q2 All three cultures get the same result if $a = 0$ or if $c = 1$.

Q3 For this problem, Terrans and Klingons get a result of 3.8, and Vulcans get 8.

Q4 Vulcans and Klingons agree on Q1, but Terrans and Klingons agree on Q2. Different cultures agree, depending on the problem.

Q5 All three calculations are the same when $c = 0$. Many other combinations of values exist.

Q6 Answers will vary, but all answers require four values. One example is $3 + 1 \cdot (-1) + 2$. The result is 4 for both Terrans and Vulcans, but not for Klingons.

Q7 Answers will vary. This question is a good jumping-off place for a class discussion. Students may notice that agreement on rule 1 (evaluate parentheses first) means that mathematicians from all three cultures can eliminate ambiguity by inserting enough parentheses to completely determine the order of operations.

Q8 Answers will vary. A lack of rules could result in many disagreements in retail transactions, in business deals, and throughout the economy, and could handicap scientists, engineers, and anyone who calculates frequently.

EXPLORE MORE

Q9 $6 - 4 \div 2 = 4$. No other culture agrees with this Terran result; both Klingons and Vulcans get 1.

Q10 $7 + 5 + 2 = 14$. All three cultures agree on this result.

Q11 $2 \cdot 4 - 3 = 5$. Klingons and Terrans agree on the result, but Vulcans get an answer of 2.

Q12 $14 \div 7 \cdot 3 = 6$. All three cultures agree on this result.

Q13 RPN is an unambiguous algebraic notation that does not use parentheses. Based on work by Polish mathematician Jan Lukasiewicz in the 1920s, it is supported by quite a few calculators, particularly from HP. Much more information is available on the web.

 ACTIVITY NOTES

Q14 On the earlier pages you performed only two operations, so there were only two possible answers. Page abcd has four variables, making it convenient to create an expression involving three operations. It's not hard to create such an expression for which all three systems get different answers. An example is $1 + 2 \cdot 3 - 4$, which comes out to 3 in Terran, -3 in Vulcan, and 5 in Klingon.

WHOLE-CLASS PRESENTATION

In this presentation students will see the results of applying different rules for the order of operations, and will gain a greater appreciation for the importance of agreeing on standard rules.

Use the Presenter Notes, along with the sketch **Order of Operations Present.gsp,** to present this activity to the whole class. As you proceed with the presentation, you will need to refer to the rules that are used in the Terran, Vulcan, and Klingon civilizations. You can see the rules at any time by pressing the *Show Rules* button, and you can return to the presentation by pressing one of the Return buttons.

You can use the custom tools to create additional examples. To get different results for all three cultures, you need to use a problem involving at least three operations. You can use page abcd to create such a problem. See Q14 for an example.

Exploring Algebra 1 with The Geometer's Sketchpad
© 2012 Key Curriculum Press

 **Presenter Notes**

In this presentation you will explore the different rules for order of operations that are observed in the Terran, Vulcan, and Klingon galactic civilizations.

Ask the class if it really makes a difference. Although the three groups use a different order, they are still performing the same operations. Won't the answers be the same? Put this question to the test by evaluating the expression $1 + 3 \cdot 2$, but using variables so we can change the numbers later.

1. Open **Order of Operations Present.gsp.** Drag the *a*, *b*, and *c* markers so that the class can see how the corresponding markers move in unison. Set up the problem by pressing the *Move Points* button.

> Press the *Show Rules* button to display the rules for each system.

Q1 Begin with the Terran system. Ask, "Which operation comes first?" In the Terran system, it is multiplication. Press the *Terran Step 1* button to show this step.

Q2 Ask, "Which operation is next?" Press *Terran Step 2* to show the addition.

Q3 Ask, "Will Vulcans get the same answer? Which operation comes first?" By Vulcan rules, the addition comes first, and multiplication second.

2. Press *Vulcan Step 1* to show $a + b$. Then press *Vulcan Step 2* to show $(a + b) \cdot c$.

Q4 Ask, "What order will Klingons use?" (Like Vulcans, Klingons will add first and then multiply.)

3. Press *Klingon Step 1* and *Klingon Step 2* to show the Klingon result.

> Since Vulcans and Klingons used the same order, their answers for this problem will match for any numbers.

Q5 Ask, "If we start with different numbers, will Terrans always differ from the others? Will Vulcans and Klingons always agree?" Drag *a*, *b*, and *c* to test student responses. The Terran answer agrees with the others if $a = 0$ or $c = 1$.

Q6 Ask, "Does this mean Vulcans and Klingons always agree, and Terrans always disagree?"

4. Go to page 2 and evaluate $12 \div 2.5 - 1$ in all three systems. This time Terrans and Klingons agree on $((a \div b) - c) = 3.8$, but Vulcans get $(a \div (b - c)) = 8$.

Q7 Again ask for values that make all three systems agree on this calculation. One solution is $c = 0$.

Discuss the reasons for having rules for the order of operations. A good comparison is the rules for driving on the right or left side of a road. Both systems are used effectively on this planet. The important thing is that everyone in a given society observes the same rule.

> The Microsoft Calculator uses Klingon when the view is set to Standard, but uses Terran when it's set to Scientific.

Ask the students if they know of any place on Earth where the Klingon rules prevail. Many inexpensive pocket calculators take operations in order, with no hierarchy. So does the Microsoft Calculator program, which is installed with Windows.

Order of Operations

In this activity you will explore the importance of agreeing on and following the standard rules for order of operations:

1. Evaluate parentheses.

2. Evaluate any exponents.

3. Perform multiplication and division from left to right.

4. Perform addition and subtraction from left to right.

Image courtesy of U.S. Postal Service

DIFFERENT STROKES FOR DIFFERENT FOLKS

All known cultures in the galaxy evaluate parentheses and exponents before performing addition, subtraction, multiplication, and division.

When you perform a series of arithmetic operations on Earth, you follow the Terran order described above. This is not the only order possible. The rules for steps 3 and 4 are different in Vulcan mathematics, and different still in Klingon mathematics.

Terran	Vulcan	Klingon
3. Multiply and divide from left to right. 4. Add and subtract from left to right.	3. Add and subtract from left to right. 4. Multiply and divide from left to right.	3. Add, subtract, multiply, and divide from left to right, in the order in which they appear.

For this reason, you, Spock (who is a Vulcan), and Worf (who is a Klingon) may all get different answers to the same problem.

1. Open **Order of Operations.gsp.**

2. On the first page, you will evaluate the expression $1 + 3 \cdot 2$ in all three systems. Think of this as $a + b \cdot c$. To prepare, drag point a to 1, b to 3, and c to 2.

On Earth, multiplication and division must be done first, from left to right, before addition and subtraction. You'll do this calculation on the Terran number line.

When you use the **Multiply** tool, click the point on the number line, not the marker.

3. On the Terran number line, you must calculate $b \cdot c$ first. Press and hold the **Custom** tools icon, choose the **Multiply** tool, and click on points b and c.

With the marker selected, choose **Display | Color.**

4. Now calculate $a + (b \cdot c)$ on the Terran number line by choosing the **Add** custom tool and clicking first on point a and then on point $(b \cdot c)$. When you finish, change the color of the resulting marker.

5. In the Vulcan system, they do addition and subtraction first, before multiplication and division. Use the **Add** and **Multiply** custom tools to do the Vulcan calculation on the second number line. Color the final result marker to match the color of the Terran answer.

Exploring Algebra 1 with The Geometer's Sketchpad
© 2012 Key Curriculum Press

6. Similarly, do the Klingon calculation and color its final result marker.

Q1 Which of the three systems get the same result for this calculation?

Q2 Drag points *a*, *b*, and *c*. Are there any values of *a*, *b*, and *c* for which all three calculations are equal? If so, what are they?

7. Go to page 2 and evaluate $12 \div 2.5 - 1$ in all three systems. Prepare by dragging point *a* to 12, *b* to 2.5, and *c* to 1. Then use the custom tools.

Q3 What answer do the various systems give for this calculation? Which ones agree?

Q4 If some systems agree, are they the same ones that agreed for the first expression you evaluated?

Q5 Are there some values of *a*, *b*, and *c* for which all three calculations are equal? If so, what are they?

Q6 Try to construct a mathematical expression for which Terrans and Vulcans agree, but Klingons get a different result. What did you find?

Q7 None of the three cultures is willing to abandon its system to adopt one of the others. In this situation, how do you think mathematicians from the three different cultures could communicate? Is it possible to create a Galactic Standard that beings from all three cultures would understand the same way?

Q8 Explain in your own words why we need rules to specify the order of arithmetic operations. What problems could arise if different people used different rules?

EXPLORE MORE

For these four questions, duplicate the Explore More page four times by choosing **File | Document Options** and using the Add Page button.

Even though the rules are different, sometimes Terrans get the same results as either Vulcans or Klingons. Perform each calculation below in all three systems. List the Terran result in the first blank column below. In the next column, list any other systems that agree with the Terran result, and in the last column, list systems that disagree.

	Calculation	Terran Result	Systems Agreeing	Systems Disagreeing
Q9	$6 - 4 \div 2$			
Q10	$7 + 5 + 2$			
Q11	$2 \cdot 4 - 3$			
Q12	$14 \div 7 \cdot 3$			

Q13 Research the RPN system and evaluate it as a possible Galactic Standard.

Q14 On page abcd, create a problem for which each system gets a different answer.

Equivalent Expressions

By manipulating and constructing algebars, students explore the commutative and distributive properties and various rules involving exponents. Even more important, by dragging variables and observing the changes in the expressions based on the variables, students get used to the dynamic behavior that variables and expressions show. This sense that algebraic expressions are changeable, that they represent an entire range of possible values, is easier for students to internalize when they can actually see the values in motion.

This activity is also valuable preparation for other activities that use algebars (for instance, Undoing Operations and Solving Linear Equations by Undoing).

INVESTIGATE

Q1 Encourage students to make a prediction before dragging. As students drag a and b, the green algebars remain the same length, because $a + b = b + a$ is an example of the commutative property of addition.

Q2 As students drag a and b, these two bars are seldom equal in length. The expressions $a - b + 1$ and $b - a + 1$ are not equivalent, except when $a = b$.

Q3 As students drag a and b, these two bars are always equal in length. The expressions are equivalent:

$$a - (b - a) = a + (a - b)$$

Q4 These bars are the same length only when $b \geq 0$. Students must drag b to the left of 0 to discover that the two expressions are not equivalent.

Q5 $ab = a\sqrt{b^2}$ when $b \geq 0$.

Q6 Answers will vary. The important thing is that students make a prediction.

Q7 The addition algebars are always the same length, and so are the multiplication algebars. The equations are $a + b = b + a$ and $ab = ba$. The subtraction and division algebars are sometimes not the same length. Therefore addition and multiplication are commutative, and subtraction and division are not.

Q8 The expressions $2(c + 4)$ and $2c + 8$ are equivalent. As an equation, $2(c + 4) = 2c + 8$. This is an example of the distributive property of multiplication over addition.

Q9 Answers will vary. Some students may describe the behavior of the bars; others may give a counter-example; and others may give an algebraic argument in terms of the distributive property. The important thing is to get students to think about the question.

Q10 Yes, $2(m + n) = (2m) + (2n)$.

Q11 No, the expressions $2 + (m \cdot n)$ and $(2 + m) \cdot (2 + n)$ are not equivalent.

Q12 Yes, $x(y - z) = xy - xz$.

EXPLORE MORE

Q13 The expressions $a^c b^c$ and $(ab)^c$ are equivalent because exponents are distributive across multiplication, but not addition.

Q14 When $c = 0$, all of the expressions are equal for any values of a and b. If c is any other value, then students must experiment with values of a and b. A simple solution would be setting $a = b = 2$ since $2 \cdot 2$ and $2 + 2$ are equal. Other numerical values for a and b make the expressions equal when $a = b/(b - 1)$.

Q15 No two of these expressions are equivalent. All three give different results when $y < 0$: the result of the first is negative, the result of the second is undefined, and the result of the third is positive.

Equivalent Expressions

In this activity you will investigate algebraic expressions that are equivalent. Equivalent expressions look different but always have the same value.

EXPRESSIONS AND ALGEBARS

Algebars are bars that represent algebraic quantities. Red bars represent variables; green bars represent expressions.

1. Open **Equivalent Expressions.gsp** and press the *Show Variables* button. Two red algebars appear representing the variables a and b.

2. Press the *Show Algebars 1* button. Two green algebars appear, labeled with their algebraic expressions: $a + b$ and $b + a$.

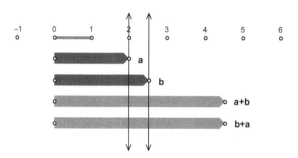

If two bars are always equal in length, they represent equivalent expressions. Try many values of the variables before you decide that two expressions are equivalent.

Q1 Predict what will happen if you change a and b by dragging the tips of the red algebars. Will the green algebars remain equal in length? Drag a and b to find out. Were you right? Are the two algebraic expressions equivalent? If so, write the result as an equation.

3. Press the *Show Algebars 2* button. Two more green algebars appear, representing two different algebraic expressions.

Use the indicator lines to estimate the values of a and b.

Q2 Drag a and b again. Are the new algebars always equal in length? Are there any positions for a and b that make these bars equal? Describe them. Are the two algebraic expressions equivalent?

4. Press the *Show Algebars 3* button. Two more green algebars appear, representing two different algebraic expressions.

Q3 Drag a and b again. Are the new algebars always equal in length? Are the two algebraic expressions equivalent? If so, write your result as an equation.

5. Press the *Show Algebars 4* button. Two more green algebars appear without their labels.

Q4 Drag a and b again. Are these bars always equal in length? If not, when are they equal? Do you think the algebraic expressions for these two bars are equivalent?

Q5 Press the *Show Expressions 4* button to see the labels for the last two bars. On your paper summarize your conclusions by filling in the blanks in a sentence like this: _____ = _____ when _____.

Exploring Algebra 1 with The Geometer's Sketchpad
© 2012 Key Curriculum Press

THE COMMUTATIVE PROPERTY

If an operation is *commutative*, you can perform it in either order (for instance, $a + b$ or $b + a$) and get the same result.

Q6 Go to page 2. This page shows four possible commutative properties (for addition, subtraction, multiplication, and division). Predict which algebars will stay the same length when you drag a and b.

Q7 Drag a and b to test your prediction. Which pairs of algebars are always the same length? Write an equation for each pair that matches. Which of the four arithmetic operations are commutative?

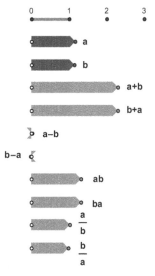

THE DISTRIBUTIVE PROPERTY

6. On page 3, drag c and note how the bars change.

Q8 Which two expressions are equivalent? Why?

Q9 Why is $2c + 4$ not equivalent to $2(c + 4)$?

Sabrina says that you can evaluate the expression $2(m + n)$ either the way it's written (add first, then multiply by 2) or by first multiplying the 2 by each of the values in the parentheses and then adding the results.

7. On page 4 are the bars Sabrina created to test her rule. Press the *Show Sabrina's Algebars* button to show them. The bottom two bars show the expression both the original way and Sabrina's way: $2(m + n)$ and $(2m) + (2n)$.

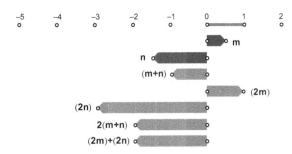

Q10 Drag m and n back and forth. Are $2(m + n)$ and $(2m) + (2n)$ equivalent? If so, write your result as an equation.

Corey says that you can do something similar with the expression $2 + (m \cdot n)$: You can first add the 2 to both values in the parentheses and then multiply the results.

8. Press the *Show Corey's Algebars* button. The last two bars show the expression both the original way and Corey's way: $2 + (m \cdot n)$ and $(2 + m) \cdot (2 + n)$.

Q11 Drag m and n. Are the expressions $2 + (m \cdot n)$ and $(2 + m) \cdot (2 + n)$ equivalent?

OTHER EXPRESSIONS

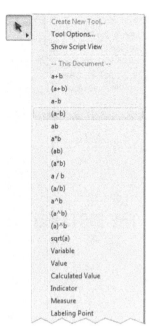

9. On page 5, build your own algebars to test whether $x(y - z) = xy - xz$. Start by constructing the expression $(y - z)$.

10. Press and hold the **Custom** tool icon and choose the **(a−b)** tool.

11. Click this tool on five objects: the top white point that's not already used, the point at the tip of the y algebar, the caption on the y algebar, the point at the tip of the z algebar, and the caption on the z algebar.

12. To finish constructing $x(y - z)$, choose the **ab** tool and click it on five objects: the starting white point, the tip and caption of the x algebar, and the tip and caption of the $(y - z)$ algebar.

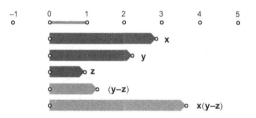

Next construct the expression $xy - xz$. Start by constructing xy and xz.

13. Construct xy by using the **ab** tool on the x and y bars. Then construct xz.

14. Construct $xy - xz$ by using the **a − b** tool on your xy and xz bars.

Q12 Drag x, y, and z. How does your test turn out? Does $x(y - z) = xy - xz$?

EXPLORE MORE

To change the label of a variable, select the point at the tip of the bar and choose **Display | Label Point.**

Q13 On page 6, build the expressions below. Use the **a^b** tool to raise a value to a power. Which of the three are equivalent? Write your answer as an equation.

$$a^c b^c \qquad (a + b)^c \qquad (ab)^c$$

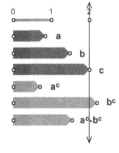

Q14 Find values of a, b, and c that make all three of these expressions equal. What values of the variables did you use? How many sets of values can you find?

Q15 On page 7, test the following expressions to see if they are equivalent. Describe your conclusions.

$$y \qquad \left(\sqrt{y}\right)^2 \qquad \sqrt{y^2}$$

Exploring Algebra 1 with The Geometer's Sketchpad
© 2012 Key Curriculum Press

Equivalent Expressions: The Border Problem

INVESTIGATE

The concept of units isn't explored explicitly in this activity. If s is the side length of the square—and is therefore measured in linear units such as centimeters (cm)—how can s also represent the area of one of the yellow rectangles, which should use square units such as cm^2? The answer is that the area of the yellow rectangle is the product of its lengths, or $(s \text{ cm}) \cdot (1 \text{ cm}) = s \text{ cm}^2$ (assuming we're in centimeters). This is a good topic for class discussion.

Q1 The area of each yellow rectangle is $1s$, or simply s. The area of all four of these is then $4s$. The area of one of the green squares is 1 since they are 1×1 squares. The area of all four squares is then 4, and the total area of the border is $4s + 4$.

Q2 Sample answers:

$$4(1) + 4 = 8$$
$$4(2) + 4 = 12$$
$$4(3) + 4 = 16$$

Q3 $4(s + 1)$

The area of any one of the four rectangles is $s + 1$ since each extends 1 unit beyond the side length s. (In other words, each is a rectangle the width of the flower bed plus a corner square.) The area of all four rectangles is then $4(s + 1)$.

Q4 Many possible answers. Here are several:

$$2(s + 2) + 2s$$
$$s + s + s + s + 1 + 1 + 1 + 1$$
$$(s + 2) + 3s + 2$$
$$(s + 2) + 2(s + 1) + s$$
$$3(s + 1) + s + 1$$

EXPLORE MORE

Q5 See Q4.

Q6 The plots are all right on top of each other.

ACTIVITY NOTES

Q7 One way to approach this would be to show that all of the expressions are equivalent to $4s + 4$. By the transitive property, they are all then equivalent to each other. To prove that $4(s + 1)$ is equivalent to $4s + 4$, for example, use the distributive property.

Q8 The area of the border and bed combined is $(s + 2)^2$, or $s^2 + 4s + 4$. Subtracting the area of the flower bed, s^2, we once again find $4s + 4$ for the area of the border.

WHOLE-CLASS PRESENTATION

This activity works well as a whole-class activity with a presentation computer and projector. Describe the situation to the class and have students make the appropriate drawing. Then ask the class for an expression for the border area. Have a student who came up with the expression justify it with a drawing. Then open up **Border Expressions.gsp** and show them this solution in Sketchpad. Calculate the area expression and ask if it seems to give the right answer for various values of *s*. Then go to page 3 and have students come up with several more expressions on their own or in groups. One by one, have students (or possibly group representatives) build their expressions in Sketchpad and do the associated calculations. Finish up the activity with a discussion of the various expressions—why some worked and some didn't.

Equivalent Expressions: The Border Problem

Suppose you want to build a square flower bed surrounded by a 1-m-wide border to serve as a path. You need to know the area of the border so that you'll know how much gravel to buy. You haven't decided how big a square the flower bed should be, but you want to be able to find the border's area for any flower bed with side length s. You need an algebraic expression.

INVESTIGATE

Equivalent expressions are expressions that look different but that give the same result, such as $x - 2$ and $x + (-2)$.

Several different equivalent expressions are possible for this problem. Your goal is to come up with as many of them as you can and test them for equivalence by seeing if they all give the same area for any side length.

1. Open **Border Expressions.gsp.**

You'll see a square with a border around it, representing the flower bed and the path. Eight quadrilateral interiors are constructed on the border—four green and four yellow. This arrangement represents one possible expression for the area of the border: $4s + 4$.

Q1 Explain how the expression $4s + 4$ relates to the quadrilaterals around the flower bed. Why does this expression give the area of the border?

If the Calculator covers the measurement of *s*, drag the Calculator aside.

2. Use Sketchpad's Calculator to find the value of $4s + 4$. To do this, choose **Calculate** from the Number menu. Then type 4, click the measurement for s in the sketch, add 4, and click **OK**.

Q2 Did you get the correct value for the area of the border? Drag the red point at the end of the s slider. Is the calculation still correct? Pick three different values of s. Show how substituting these values for s gives the correct value for the area of the border.

3. Go to page 2. You'll see another arrangement of quadrilaterals around the border.

Q3 What expression does this arrangement represent?

s = 3.00

flower bed

s

4. Again, use Sketchpad's Calculator to find the value of your expression. Do you get the correct value for the area even when you drag the red point?

5. Go to page 3. You'll see flower beds for the two previous arrangements, and many more with borders yet to be filled in.

To construct a quadrilateral interior, select its four vertices in order and choose **Construct | Quadrilateral Interior.**

6. On one of the empty flower beds, construct a different set of quadrilaterals that divide the border area in a new way to represent a different expression for the area.

7. Use Sketchpad's Calculator to calculate the value of your expression. Does your expression work even as you drag the red point to change the value of *s*?

8. Repeat steps 6 and 7 two more times.

Q4 Write your three new equivalent expressions from steps 6–8.

EXPLORE MORE

Q5 Find as many more equivalent expressions as you can.

Q6 Plot the functions associated with all of your equivalent expressions. How do the plots demonstrate their equivalence?

(For the expression $4s + 4$, for example, plot the function $f(x) = 4x + 4$. To do this, go to page 4, choose **Graph | Plot New Function,** enter $4x + 4$, and click **OK.**)

Q7 Use properties of algebra to prove that all of your expressions are equivalent.

Q8 Find an equivalent expression by calculating the area of the bed and subtracting it from the area of the border and bed combined. Simplify your result and show that it is equivalent to your other expressions.

To use this activity with measurements in meters rather than feet, use pages "1(m)" and "2(m)", and use the **Scaled Length (m)** tool.

SKETCH AND INVESTIGATE

Q1 The result for Cari's method is

$$(25 \text{ ft})(80 \text{ ft}) + (25 \text{ ft})(100 \text{ ft})$$

$$= 2000 \text{ sq ft} + 2500 \text{ sq ft}$$

$$= 4500 \text{ sq ft}$$

Q2 The result for Zeeba's method is

$$(25 \text{ ft})(80 \text{ ft} + 100 \text{ ft})$$

$$= (25 \text{ ft})(180 \text{ ft})$$

$$= 4500 \text{ sq ft}$$

Q3 $ab + ac = a(b + c)$

Q4 Both methods still give the same result. For these particular numbers, the result is 3500 sq ft.

Q5 In the case of three walls, the factor a would be multiplied by the widths of all three walls:

$$ab + ab + ac = a(b + b + c) = a(2b + c)$$

THREE WALLS

Q6 Both methods do give the same result: 6500 sq ft.

Q7 See Q5.

Q8 Answers will vary. In the perspective view the scale is variable. This is because objects are foreshortened, and some objects are closer to the viewer than others. Moreover, a rectangle generally does not even appear as a rectangle.

WHOLE-CLASS PRESENTATION

Use **Distributive Painting Present.gsp** in conjunction with the Presenter Notes to present this activity to the whole class.

The Distributive Property: A Painting Dilemma

1. Open **Distributive Painting Present.gsp,** and explain the problem to your class based on the description at the beginning of the student activity sheet.

2. Press the *Cari's Solution* button. Explain that this is the configuration on which Cari decided to base her calculation.

3. Explain that the first part of Cari's method is to calculate the area of the rectangle on the left. Choose **Number | Calculate** to show the Calculator. Click the 25 ft measurement in the sketch, the multiplication sign on the keypad, and the 80 ft measurement in the sketch. Click **OK** to finish the calculation.

4. Use a similar calculation to compute the area of the rectangle on the right. Then use the Calculator one more time to find the sum.

Q1 Ask, "How would you express this calculation using the variables *a*, *b*, and *c*?" (The label of the calculation gives the answer away. Congratulate students for noticing this.)

Now calculate the area using Zeeba's method.

5. Press the *Reset* button to return the walls to their original arrangement. Press the *Zeeba's Solution* button.

6. Use the Calculator to multiply the height (25 ft) by the length (180 ft). Be sure to click the measurements in the sketch; don't type the numbers on the keypad.

Q2 Ask, "How would you express this calculation using the variables *a*, *b*, and *c*?" (The label of the calculation again gives the answer away. Don't congratulate the students this time.)

Q3 Ask, "Are the results equal?"

Q4 Ask, "Will the results always be equal, even if the dimensions are different?"

Even if the entire class says the results will always be equal, tell them that it's too easy to be fooled in problems like this and that it's worth trying different measurements.

7. Press the *Show Dimensions* button, double-click the *height* measurement, and change it to 20. Double-click the *height* measurement and change it to 60. Press *Reset*.

8. Press the *Cari's Solution* button to show Cari's result, and then the *Zeeba's Solution* button to show Zeeba's.

Q5 Ask students to summarize the principle of algebra that they've found by investigating this problem.

9. If there's time, explore page 2 (with three walls).

The Distributive Property: A Painting Dilemma

The school activities committee is preparing to paint two gymnasium walls. Both walls are 25 ft high. The first wall is 80 ft wide, and the other is 100 ft wide.

Cari and Zeeba are on the committee, and they volunteered to calculate how many cans of paint to order. To figure this out, they need to find the total area of the walls. But Cari and Zeeba disagree about how to do the calculation.

Cari wants to calculate the area of each wall separately and then add them together to get the total painted area:

$$(25 \text{ ft})(80 \text{ ft}) + (25 \text{ ft})(100 \text{ ft})$$

Zeeba wants to calculate the total area by first adding up the widths of the walls to get the total width and then multiplying the height by this total:

$$(25 \text{ ft})(80 \text{ ft} + 100 \text{ ft})$$

SKETCH AND INVESTIGATE

1. Open **Distributive Painting.gsp.** This is a perspective view of the two walls as seen from inside the gym.

First calculate the area using Cari's method.

2. Press the *Cari's Solution* button.

3. To measure the walls, press and hold the **Custom** tool icon and choose **Scaled Length.** Click this tool on line segments *a*, *b*, and *c*.

4. To calculate the area using Cari's method, first choose **Number | Calculate** and find the area of the rectangle on the left. To enter the width and height of this wall into the Calculator, click the measurements in the sketch.

If you prefer, you can do all three calculations in one step.

5. Calculate the area of the rectangle on the right. Then finish calculating by Cari's method by using the Calculator one more time to find the sum of the two areas.

Q1 What is the result using Cari's method?

Now calculate the area using Zeeba's method.

6. Press the *Reset* button to return the walls to their original arrangement. Press the *Zeeba's Solution* button.

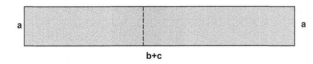

7. Measure the line segment labeled $b + c$.

8. Calculate the area using Zeeba's method, by using Sketchpad's Calculator to calculate the total area of this single rectangle.

Q2 What is the result using Zeeba's method?

Q3 Write your expressions as an equation using the variables *a, b,* and *c*. This equation is a symbolic statement of the *distributive property of multiplication over addition.*

You've tried your result for only one set of measurements. Jason claims that it might work for this one case, but it won't work for others.

To change the length, double-click it and type a new value.

9. To test Jason's claim, press the *Show Dimensions* button, change the length from 100 ft to 60 ft, and press *Reset.* Then press the buttons to try both Cari's solution and Zeeba's solution.

Q4 Do both methods still give the same result?

Q5 Make a conjecture. How would the distributive property apply if there were three walls, two with width *b* and one with width *c*?

THREE WALLS

10. Go to page 2 of the document.

11. After calculating the total area of the walls, the students find that they have enough money to paint a third wall in the gym. Repeat the previous steps and calculate the painted area using both Cari's and Zeeba's methods.

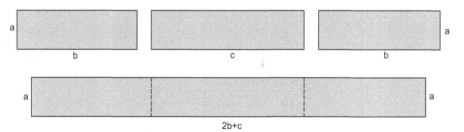

Q6 Do both methods give the same result? What is the combined area of the three walls?

Q7 Write your expression as an equation. Compare the equation with your conjecture in Q5.

Q8 It would be possible to measure the line segments while the sketch is showing a perspective view, but the results would not be very useful. Explain why this is true.

Exploring Algebra 1 with The Geometer's Sketchpad
© 2012 Key Curriculum Press

The Distributive Property

DISTRIBUTIVE PROPERTY OF MULTIPLICATION OVER ADDITION

Q1 The left side is done in two steps: (a) Add 3 + 1 to get 4, and (b) multiply 2 · 4 to get 8. The right side requires three steps because multiplying by 2 is distributed over the other two numbers: (a) Multiply 2 · 3 to get 6, (b) multiply 2 · 1 to get 2, and (c) add 6 + 2 to get 8.

Q2 The animation for the very first example shows that either way of doing the problem results in 2 blocks of 1 and 2 blocks of 3, even though they are arranged differently. When these blocks are combined into an answer, the same blocks are being combined, so the result must be the same.

Q3 Even when one of the numbers being added is negative, the distributive property is still true. You just have to watch the blocks more carefully because some of the blocks are positive and some are negative, causing them to overlap.

Q4 Answers will vary.

EXPLORE MORE: OTHER DISTRIBUTIVE PROPERTIES

Q5 The formula becomes

$$a \cdot (b + c) = (a \cdot b) + (a \cdot c)$$

This is the formula for the distributive property of multiplication over addition. The bars representing the left and right sides will always be of equal length.

Q6 The formula becomes

$$a + (b \cdot c) = (a + b) \cdot (a + c)$$

This formula represents distributing addition over multiplication. The bars don't always match, so this formula is false in general (although students may find some specific numeric values for which it is true).

Q7 The formula on page 4 is the distributive property of division over addition:

$$a/(b + c) = (a/b) + (a/c)$$

The bars do not match, and this formula is false.

The formula on page 5 is similar, except that the division sign is on the right rather than the left:

$$(a + b)/c = (a/c) + (b/c)$$

These bars always match, and the formula is true.

Discuss why the formula on page 4 (called the left distributive property of division over addition) is false, but the one on page 5 (the right distributive property of division over addition) is true.

If students don't bring it up in the discussion, ask whether there's any relationship between the formula on page 5 and the formula they explored on page 2. The connection involves using multiplicative inverses and the commutative property of multiplication, as shown here:

$$(a + b)/c = a/c + b/c$$
$$(a + b) \cdot \frac{1}{c} = a \cdot \frac{1}{c} + b \cdot \frac{1}{c}$$
$$\frac{1}{c} \cdot (a + b) = \frac{1}{c} \cdot a + \frac{1}{c} \cdot b$$

WHOLE-CLASS PRESENTATION

Use page 1 of **Distributive Property.gsp** to demonstrate the property and stimulate a discussion as to why the two expressions are equal. Use the remaining pages to investigate other possible distributive properties.

Exploring Algebra 1 with The Geometer's Sketchpad
© 2012 Key Curriculum Press

The Distributive Property

In the expression $a(b + c)$, the parentheses indicate that you must add the b and c first, and then multiply a by the result. But the distributive property says that you can distribute the a to the values in the parentheses, multiplying a by b and also multiplying a by c, and then adding the two results together.

DISTRIBUTIVE PROPERTY OF MULTIPLICATION OVER ADDITION

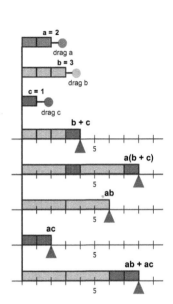

1. Open **Distributive Property.gsp.** Click the *Present Left Side* and *Present Right Side* buttons to show the two ways of calculating $a(b + c)$.

Q1 Write down the results of each step of the calculation.

2. Reset the sketch, drag the variables to represent $-3(1 + 4)$, and use the buttons to show the calculations again. Then try $2(-3 + (-1))$.

Q2 Explain in your own words how the animation demonstrates the distributive property and why the distributive property is true.

Q3 Drag the variables to represent $2(5 + (-3))$. Describe what happens when the rectangles overlap. Is the distributive property still true in this case?

Q4 Make up three more problems of your own. Use the sketch to show them, and write down the problems, the results, and anything interesting you observe.

EXPLORE MORE: OTHER DISTRIBUTIVE PROPERTIES

You can write the distributive property in a more general way, representing the two operations by $\Diamond$ and $\otimes$:

$$a \Diamond (b \otimes c) = (a \Diamond b) \otimes (a \Diamond c)$$

The last page of the sketch has directions for using these custom tools.

Q5 Rewrite this formula replacing $\Diamond$ with multiplication and $\otimes$ with addition. This is the normal distributive property you explored above. On page 2 use the **Add, Subtract, Multiply,** and **Divide** custom tools to calculate both sides of the formula and verify that the resulting bars are the same length.

Q6 Rewrite the formula replacing $\Diamond$ with addition and $\otimes$ with multiplication. Use page 3 to determine whether this modified distributive property is true.

Q7 Additional modified distributive properties are shown on pages 4 and 5. Verify or disprove each of them by constructing and comparing bars.

Algebra Tiles

You may consider creating your own activities by using the custom tools in **Algebra Tile Tools.gsp.**

THE TILE TOOLS

If students don't already have experience using custom tools, introduce the activity by showing them how to press and hold the **Custom** tool icon, how to choose the **1** tool, and how to click in the sketch to create unit squares. But don't show them the results of using the **x** tool. The behavior of the **x** tool is the focus of step 3 and Q1; let them experiment and make their own observations.

Q1 Here are some observations students may make: The x tiles are rectangular and blue; some come out horizontal and some vertical. Adjusting the x slider changes the tile size to match the slider. Dragging the black point changes a tile between horizontal and vertical.

Q2 The train represents the value 3.

5. Some students may need coaching as they try to connect and orient tiles. It may be useful to point out that they need to connect the tiles first, and that they can always flip them later. It may also be useful to observe that students must deselect a new tile before flipping it.

Q3 To model $2x + 3$, construct two x segments and three unit segments end-to-end. To construct $x + 3y + 1$, do the same with one x segment, three y segments, and one unit segment. Use the flippers to orient the trains correctly.

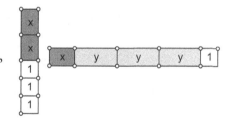

Q4 Students may make some of these observations: The x and x^2 tiles have the same color, and so do the y and y^2 tiles. The xy tiles are intermediate in color between the x and y tiles. The unit, x^2, and y^2 tiles are square, and the others are rectangular. Only rectangular tiles have flippers. The x, x^2, and xy tiles share their x dimension, and the y, y^2, and xy tiles share their y dimension.

Q5 The rectangular tiles have flippers to change them between vertical and horizontal. For square tiles, a flipper would have no effect.

FRAMING THE TILES

Q6 The width of this rectangle is $3x$, and the height is 3.

Q7 For this rectangle, you need only x tiles, and you need 9 of them. This corresponds to the multiplication $3 \cdot 3x = 9x$.

Q8 The tiles that fill the rectangle provide a way of describing its area, so we can say the area is $9x$.

Q9 Here are the answers for pages 3, 4, and 5:

3: 1 y^2 tile, 4 y tiles, and 3 unit tiles

4: 2 x^2 tiles, 7 x tiles, and 6 unit tiles

5: 2 y^2 tiles, 2 xy tiles, 3 x tiles, 5 y tiles, and 3 unit tiles

EXPRESSIONS FOR AREA

Q10 The area of the rectangle is $y^2 + 2xy + 3y + 4x + 2$.

Q11 The area formula gives $(y + 2x + 1)(y + 2)$.

Q12 $(y + 2x + 1)(y + 2) = y^2 + 2xy + 3y + 4x + 2$

Q13 Students may not have time to finish all these pages. Here are the answers for pages 7–10:

7: $(3x + 2)(y + 4) = 3xy + 12x + 2y + 8$

8: $(x + 2y + 3)(y + x) = x^2 + 2y^2 + 3xy + 3x + 3y$

9: $(4x + 1)(y + 2 + x) = 4x^2 + 4xy + 9x + y + 2$

10: $(x + y + 1)(2 + x + 3y) = x^2 + 3y^2 + 4xy + 3x + 5y + 3$

WHOLE-CLASS PRESENTATION

To present this activity to the whole class, use the Presenter Notes and the sketch **Algebra Tiles Present.gsp.**

Algebra Tiles

In this presentation you'll show your class how dynamic algebra tiles work and use algebra tile diagrams to write two different expressions for the same area.

1. Open **Algebra Tiles Present.gsp.** Page 1 is empty except for sliders representing x and y. These sliders determine the dimensions of the tiles.

2. Press and hold the **Custom** tool icon to display the Custom Tools menu. Look at the list of tools and ask students to guess what the various tools might do.

3. Choose the **1** tool and click several times on the screen. First construct several disconnected unit tiles, and then connect three together to make a train.

Q1 Use the **Arrow** tool to drag the train. Ask what number the train represents. [3]

4. Choose the **x** tool and click several times to construct disconnected tiles. Make enough that some come out horizontal and some vertical.

Q2 Ask, "What do you observe? How are the x tiles and the unit tiles different?"

5. Use the **Arrow** tool to drag the flipper point and orient the x tiles horizontally or vertically. Adjust the slider to change the value of x.

After you finish connecting the train, you may need to adjust the orientation of some of the tiles.

Q3 On page 2, use the **1** tool to create a train of three tiles on the vertical axis. Ask, "What number does this train represent?" [3]

Q4 Use the **x** tool to create a train of two tiles on the horizontal axis. Ask, "What quantity does this train represent?" [$2x$]

Q5 Ask, "How do these two trains define a rectangle?" Solicit answers from several students. Ask, "What's its width? What's its height?" [$w = 2x, h = 3$]

6. Fill in the rectangle using the **x** tool. Ask, "What does filling in the rectangle tell us about its area?" [$A = 6x$]

7. Page 3 already has horizontal and vertical trains. Ask students, "What would be the width and height of this rectangle?" [$w = 2y + 1, h = y + 3$]

Q6 Ask, "If we know the width and height of a rectangle, how could we find the area?" [Multiply $w \cdot h$] After students respond, click *Show Formula*.

8. Say, "This is a complicated multiplication. Instead, let's fill it in." Use the tile tools to fill the rectangle.

Q7 Ask, "What's the area of this rectangle?" After students answer, click *Show Area*.

Q8 Ask, "How are the multiplication formula and the number of tiles in the rectangle related?" [The number of tiles gives the result of the multiplication.]

9. Do the similar problems on the remaining pages.

Q9 Ask students to summarize their conclusions about the two ways of representing the areas of the rectangles.

Algebra Tiles

In the development of mathematics, geometry came long before algebra. Scholars tended to value geometry because they could see it and draw it. Numbers by themselves were hardly more than rumors. Today we use numbers (including variables) much more freely, but do not dismiss the lessons of the ancients. Geometry still provides us a means to lay the numbers out before our eyes. In this activity you will use algebra tiles to model numbers as lengths and areas.

THE TILE TOOLS

1. Open **Algebra Tiles.gsp.**

2. Press and hold the **Custom** tool icon to display the Custom Tools menu. Choose the **1** tool. Click several times in empty space on the screen. The tiles that appear are called *unit* tiles.

3. Choose the **x** tool and click several times. Then choose the **Arrow** tool.

Q1 Describe the *x* tiles. Drag the *x* slider on the left side of the sketch. What effect does this have? Drag the black point of each *x* tile. What happens?

4. Choose the **1** tool again. Construct a horizontal train of three unit tiles. Construct the first tile in empty space. Attach the second tile to the upper-right vertex of the first, and attach a third tile to the second.

Q2 What number does this train represent?

You can adjust each tile as you construct it, or you can first make the whole train and then adjust any tiles that come out the wrong way.

5. Construct a vertical train of four *x* tiles using the **x** Custom tool. Attach each tile after the first one to the lower-left vertex of the previous tile. If any tiles come out horizontal, use the **Arrow** tool to make them vertical.

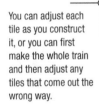

6. Adjust the *x* slider with the **Arrow** tool to make sure that the tiles stay attached as they change size. If they don't, use **Edit | Undo** to remove the tiles that don't stick together, and then use the **x** tool to redo the construction.

Q3 Use the tools to construct a horizontal train of length $2x + 3$. Construct a vertical train of length $x + 3y + 1$. Check your trains by adjusting both the *x* and the *y* slider. Sketch and describe the images on paper.

7. Try each of the remaining custom tools. Use the **Arrow** tool to adjust the black points of any tiles that have them.

Q4 Describe the similarities and differences between the tiles on your screen.

Q5 The black points that appear on some tiles are called flippers. Why do you think some tiles have flippers and others do not?

FRAMING THE TILES

The rectangle is shown here in light gray, but does not appear in the sketch.

Q6 On page 2 is a frame with two trains. What is the width of the imaginary rectangle defined by the trains? What is the height?

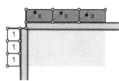

8. Start from the white point at the upper-left corner inside the frame and use tiles to completely fill the rectangle defined by the trains.

9. Once you finish, adjust the *x* slider with the **Arrow** tool to make sure that your tiles stay attached and that they fill the rectangle correctly.

Q7 What kind of tiles did you use? How many of them did you need?

Q8 How do the tiles that fill the rectangle relate to the area of the rectangle?

10. Try out the *Hide/Show Flippers* and the *Hide/Show Labels* buttons. After you observe what they do, leave the flippers and the labels showing.

Q9 On pages 3, 4, and 5 are trains that define more imaginary rectangles. On each page, fill the rectangle with tiles and write down how many of each tile you used.

EXPRESSIONS FOR AREA

Now you'll explore two ways of using tiles to find the area of a rectangle.

11. On page 6 construct trains to show a width of $y + 2x + 1$ and a height of $y + 2$.

12. Tile the rectangle. Drag *x* and *y* to check your work.

Q10 One way to find the rectangle's area is to count the tiles used to fill it. Write an expression to add up the tiles.

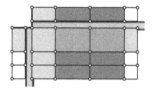

Q11 Another way to find the rectangle's area is to multiply its width by its height. Write such an expression for this rectangle.

Q12 These two ways of finding the area must be equal. Write an algebraic equation showing the multiplication (width times height) on one side and the sum of the tiles on the other side.

Q13 On pages 7 through 10 are more empty frames. On each page, construct trains matching the given width and height, and fill the imaginary rectangle with tiles. Drag *x* and *y* to test your construction. Then write down the two expressions for the area, one showing the width times the height and one showing the number of tiles that fill the rectangle.

Exploring Algebra 1 with The Geometer's Sketchpad
© 2012 Key Curriculum Press

The Product of Two Binomials

It may take the students a few minutes to get the hang of placing new tiles. When using the custom tools to create tiles, it is very useful to click on an existing point in the sketch (as opposed to clicking in blank space or on some other type of object). This serves to anchor the new tile. The point you want to click on will be highlighted when the tool is positioned properly; when you see the point highlighted, it's time to click. If you make a mistake, just choose **Edit | Undo.** Improperly constructed tiles will not stay in alignment when you adjust the sliders.

It's important that students drag the sliders for x and y periodically—this is the big advantage of using Sketchpad algebra tiles, after all. First, dragging tests whether they've used the custom tools properly. But more importantly, it reinforces the fact that x and y are variables and that the relationships discovered work no matter what their values.

INVESTIGATE

2. When students use a custom tool to construct a new tile, the tile is anchored where they click, but its direction is not yet determined as horizontal or vertical. To orient the tile vertically or horizontally, students must use the **Arrow** tool to drag the black point on the arc near the upper-left corner of the tile.

   Some students may prefer to orient each tile as soon as they construct it. Others may prefer to attach several tiles at various angles and then orient them all later.

Q1 The term x^2 refers to the big square in the upper left of the rectangle you built inside the frame. It's the product of the x from $(x + 3)$ and the x from $(x + 5)$.

The term $8x$ refers to all of the non-square tiles: $5x$ from the lower left and $3x$ from the upper right. The $5x$ is the product of the x from $(x + 3)$ and the 5 from $(x + 5)$. The $3x$ is the product of the x from $(x + 5)$ and the 3 from $(x + 3)$.

The number 15 refers to the 15 unit squares at the lower right of the rectangle. It's the product of the 3 from $(x + 3)$ and the 5 from $(x + 5)$.

Q2 $y^2 + 5y + 4$

Q3 a. $x^2 + 5x + 6$

b. $2y^2 + 7y + 3$

c. $x^2 + xy + 2x + 2y$

d. $x^2 + 4x + 4$

e. $2x^2 + 5xy + 2y^2$

f. $6y^2 + 8y + 2$

EXPLORE MORE

Q4 Possible answers: Use an "opposite" color to represent negatives, or use dashed lines or shading. Students will come to appreciate the difficulty of representing negatives (particularly negative areas).

The Product of Two Binomials

For GSP5

Mono-, bi-, and *tri-* are prefixes from the Greek words for one, two, and three, respectively.

The expression $x + 3$ is called a *binomial* because it consists of two *monomial* terms: x and 3. The expression $(x + 3)(x + 5)$ is the product of two binomials, $x + 3$ and $x + 5$. In this activity you'll use Sketchpad algebra tiles to model expressions equivalent to the products of binomials. The process you'll learn, called *expanding,* is used for writing expressions in different forms and for demonstrating the equivalence of algebraic expressions.

INVESTIGATE

1. Open **Binomial Product.gsp.**

You'll see the factored expression $(x + 3)(x + 5)$ modeled with algebra tiles. The blue tiles represent x and the yellow tiles represent one unit. Notice that one binomial factor is modeled as a vertical train and the other is modeled as a horizontal train.

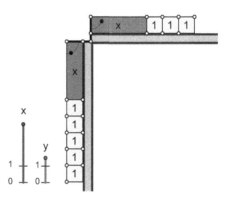

Press and hold the **Custom** tool icon to see the available tools. Choose the tool you want to use; then click a point in the sketch to construct that tile. You can drag the black point later to rotate the shape to the orientation you want.

2. Use the custom tools that come with the sketch to tile the rectangle, using tiles whose dimensions match the horizontal and vertical trains.

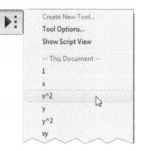

When you're done, adjust the x slider and see if your rectangle holds together.

The expression you modeled is $x^2 + 5x + 3x + 15$. You can combine the like terms to get $x^2 + 8x + 15$. This *trinomial* (an expression with three monomial terms) is called the *expanded form* of $(x + 3)(x + 5)$.

Q1 Explain how each of the terms in the trinomial $x^2 + 8x + 15$ is related to the product of the binomials.

3. Press *Hide/Show Values* to show the values of x and $y.$

Choose **Number | Calculate** to open the Calculator. Click the *x* measurement in the sketch, and type from your keyboard to build the expressions.

4. Use Sketchpad's Calculator and the measurement for *x* to calculate the values of the expressions $(x + 3)(x + 5)$ and $x^2 + 8x + 15$ for the current value of *x*.

5. Change the length of the *x* slider to confirm that the expressions remain equivalent for different values of *x*.

6. Go to page 2. Use the **y** and **1** custom tools to build trains representing the product $(y + 1)(y + 4)$. Attach each tile to a white corner of the frame or the previous tile.

To orient the tiles horizontally or vertically, drag the black arc points with the **Arrow** tool.

7. Orient the tiles horizontally or vertically, depending on where they are used.

8. Use the various custom tools to add tiles within the frame. Drag the *y* slider to make sure your tiling correctly represents this area.

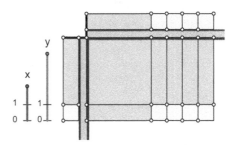

Q2 Write the expanded expression represented by the rectangle. Combine like terms.

9. Press *Hide/Show Values* to show the values of *x* and *y*. Then use Sketchpad's Calculator and the measurement for *y* to calculate the values of the expression $(y + 1)(y + 4)$ and the expanded expression you just found. Adjust the *y* slider to confirm that these expressions are always equivalent.

Q3 Build and expand the following expressions on the remaining pages of the sketch. Draw the models on paper. Write each expression both as a product of binomials and as a trinomial.

a. $(x + 2)(x + 3)$ b. $(2y + 1)(y + 3)$ c. $(x + y)(x + 2)$

d. $(x + 2)(x + 2)$ e. $(2x + y)(x + 2y)$ f. $(3y + 1)(2y + 2)$

EXPLORE MORE

Q4 Experiment with ways the model can be altered to represent expressions with negatives. For example, how could you represent $(x + 2)(x - 3)$? Illustrate and explain any models you think of.

Squaring Binomials

TILING THE SQUARE

Q1 The x and y sliders change the sides of the square.

Q2 The side length of this square is x. (Students can test this by dragging it next to the x slider.) The area is x^2.

Q3 The side length of this square is y. (Students can test this by dragging it next to the y slider.) The area is y^2. The sum of the areas of these two squares is $x^2 + y^2$.

Q4 Generally, this is not possible. There is extra space in the outlined square that cannot be covered by the two smaller tiles. Some students may discover a special case. It is, in fact, possible if x or y is equal to zero. They can model this case by changing the control sliders.

Q5 These tiles will fill the square no matter what lengths are used for x and y. Two possible solutions are shown below:

 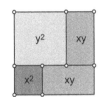

Q6 Below is one pattern that will work:

$(4x)^2 = 16x^2$ $2(4x)(y) = 8xy$ y^2

The equation is $(4x + y)^2 = 16x^2 + 8xy + y^2$.

BINOMIALS WITH SUBTRACTION

Q7 If the squares are covered as shown below, the only light color remaining will be a square with area $(x - y)^2$.

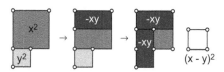

WHOLE-CLASS PRESENTATION

To present this activity to the whole class, use the Presenter Notes and the sketch **Squaring Binomials Present.gsp.**

Exploring Algebra 1 with The Geometer's Sketchpad
© 2012 Key Curriculum Press

Squaring Binomials

In this presentation students will see a visual representation of the squaring of a binomial, and will make a connection between the squares and rectangles on the screen and the various terms that make up the algebraic expression.

1. Open **Squaring Binomials Present.gsp.** The empty square represents the square of the binomial $x + y$.

2. Change the value of x by dragging the point at the end of the blue slider.

Q1 Ask, "What happens when the value of x changes? How does it affect the two colored squares? How does it affect the rectangles?" (Only one square, representing x^2, is affected, and only the x dimension of the rectangle changes.)

> Be sure to leave the values of x and y significantly different.

3. Similarly, change the value of y and have students observe the effects.

4. Drag the colored shapes into the empty square and arrange them so they fill the empty square. To flip a rectangle to the correct angle, drag its black point.

Q2 Ask, "Geometrically, the shapes fit exactly. What does this mean algebraically?" Have students describe the connection between the shapes and the four terms on the right-hand side of $(x + y)^2 = x^2 + xy + xy + y^2$.

> If you use the custom tools to attach the shapes, they will fill the square even when you change x and y.

5. Drag x or y to change the size of the empty square. The pieces are no longer in the correct positions. Press the *Tile* button to move the shapes so they fit again. Make sure students are convinced that the shapes will always fit.

6. Go to page 2, representing $(4x + y)^2$. Drag x and y to change the shapes. There are lots of tiles to move here, so use the *Tile* button right away.

Q3 Ask students to count the shapes and to write a formula based on the number of squares and rectangles they count: $(4x + y)^2 = 16x^2 + 8xy + y^2$.

> The dark rectangles can only be used to cover up positive area.

7. Go to page 3. This page represents $(x - y)^2$. Drag x and y to show how the shapes change. Point out that the dark $-xy$ rectangles represent negative area.

8. Add the x^2 and y^2 tiles by placing them next to each other such that one vertex and one side coincide. Drag the $-xy$ rectangles so they cover as much of the positive area as possible.

Q4 Ask, "How much positive area is left after the subtraction?" Students may guess that the remaining positive area is equal to the $(x - y)^2$ square. Confirm this by dragging the $(x - y)^2$ square so it coincides with the remaining positive area.

9. Change the values of x and y, switching which of them is larger. Separate the tiles, and then use the *Tile* button to arrange them again.

Q5 Ask students to write the formula that the tiles illustrate:
$(x - y)^2 = x^2 + y^2 - xy - xy$.

Squaring Binomials

Ask an algebra teacher what the most common mistakes are among algebra students. This one is certain to be near the top of the list:

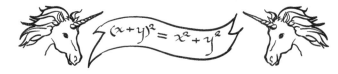

In general, this equation is not correct. There is no property that allows you to distribute exponents over addition this way. Yet even the best students continue to make this mistake years after they have learned better. The fallacy of this equation becomes clearer when you model it geometrically with the help of algebra tiles.

TILING THE SQUARE

1. Open **Squaring Binomials.gsp.**

Q1 Drag the x and y sliders. What effect do they have on the square?

Use the **Arrow** tool to drag the endpoints of the sliders.

The sides of the square are $x + y$, so its area must be $(x + y)^2$.

2. Press and hold the **Custom** tool icon. Choose the **x^2** tool from the menu that appears. Click in an open place on the screen to create the square.

Q2 In terms of x and y, how long is each side of the new square? What is its area?

3. Choose the **y^2** custom tool and use it to construct another square in open space.

Q3 In terms of x and y, how long is each side of this square? What is its area? What is the sum of the areas of the two squares you just constructed?

If the equation $(x + y)^2 = x^2 + y^2$ is correct, then the area of the outlined square (on the left side of the equation) must be equal to the sum of the areas of the two new squares you just constructed (on the right side of the equation).

Q4 Test whether the equation is true by dragging the two new squares into position so that they precisely cover the outlined square. Can you do this? Explain your answer.

Each xy has one black point. Drag this point to flip the rectangle.

4. In fact, $(x + y)^2 = x^2 + 2xy + y^2$, so you will need more tiles to fill the square. Choose the **xy** custom tool. Click twice on the screen. Choose the **Arrow** tool again.

Q5 Is it possible to fill the large square with the tiles you have created? Move the tiles into place to fill it. Draw a diagram showing how you did this.

Exploring Algebra 1 with The Geometer's Sketchpad
© 2012 Key Curriculum Press

What if you square a more complicated binomial? Consider $(3x + 2y)^2$. You can see four parts of the square separated in the image on the right. By counting shapes, you can see that $(3x + 2y)^2 = 9x^2 + 12xy + 4y^2$.

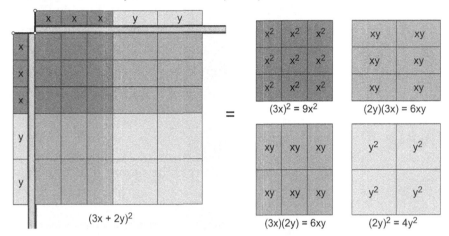

Q6 Now try it with a different binomial. Go to page 2. There is a square representing $(4x + y)^2$. Use the custom tools to form four groups of tiles, as shown in the preceding example. Then drag the tiles into the large square to fill it. Draw a diagram of the results and write the equation they represent.

BINOMIALS WITH SUBTRACTION

What if the binomial has a subtraction sign separating the two terms? In that case, the middle term of the expansion is negative:

$$(x - y)^2 = x^2 - 2xy + y^2$$

In order to show this geometrically, you need a way to show negative area.

The dark rectangles represent negative area.

5. Go to page 3 of the document. There are light-colored squares and two dark rectangles.

To add area, put two shapes next to each other. To subtract area, cover up a positive area with a negative area.

6. Add x^2 and y^2 by putting their squares next to each other. Then subtract $2xy$ by covering as much of the positive area as you can with the $-xy$ rectangles.

Q7 How much positive area is left? How does this remaining area compare with the $(x - y)^2$ square? Draw a diagram of your results from step 6.

An important concept that this activity touches on but does not cover in depth is rational versus irrational numbers. You may want to start the activity by giving students a little background on this topic. Interesting questions to ask (once the definitions are understood) are: "Could the square of an irrational number be a rational number?" "Could the square root of an irrational number be a rational number?" and "Could a square have an irrational side length but a rational area?"

INVESTIGATE

Q1 The area of a square is the *square* of its side length.

The side length of a square is the *square root* of its area.

(This is why taking something to the second power is called "squaring" and why "square root" is called what it is.)

Q2 The formula for the area of a square is $A = s \cdot s$ (where s refers to the length of a side). Multiplying something by itself *is* squaring; thus, area is the square of side length. Similarly, finding out what number times itself gives a certain result is finding the square root; thus, side length is the square root of area.

Q3 There are *exactly* 12 whole numbers less than or equal to 20 whose areas can be represented on a grid, or 13 if you include 0. The completed table should contain 12 of the 13 entries listed here.

n	0	1	2	4	5	8	9	10	13	16	17	18	20
$\sqrt{n}$	0	1	1.41	2	2.24	2.28	3	3.16	3.61	4	4.12	4.24	4.47

Q4 No; students will have seen that some whole numbers can't be modeled as the areas of squares on a grid. Additional trial and error will show that some areas, such as 7, cannot be modeled with a square on this grid.

PATTERNS

Q5 2, 8, 18, 32, 50, 72, 98, 128, ... Dividing each term by 2 will reveal the sequence of perfect squares. The rule is $2n^2$ where n is the term number. Some students may see instead the pattern of differences between terms: 6, 10, 14, 18, and so forth.

Q6 Students may list any three of these answers; each is the third number of a Pythagorean triple:

5 (3 − 4 − 5) 15 (9 − 12 − 15)

10 (6 − 8 − 10) 17 (8 − 15 − 17)

13 (5 − 12 − 13) 20 (12 − 16 − 20)

EXPLORE MORE

Q7 Different rearrangements are possible, but one way is to use the diagonals to divide the square and rearrange the four resulting triangles into two squares. (This also illustrates why the area rule is $2n^2$, since the area of each resulting square is n^2.)

Q8 $\sqrt{2}, \sqrt{8}, \sqrt{18}, \sqrt{32}, \sqrt{50}, \sqrt{72}, \sqrt{98}, \sqrt{128}, \ldots$ which can be simplified to $\sqrt{2}, 2\sqrt{2}, 3\sqrt{2}, 4\sqrt{2}, 5\sqrt{2}, 6\sqrt{2}, 7\sqrt{2}, 8\sqrt{2}, \ldots$. The numeric rule is $n\sqrt{2}$, which is the square root of $2n^2$. This makes sense because the side length of a square is the square root of the area.

Q9 Answers will vary. Sample answer: Start with an unslanted square of area 9 sq. units and with the red vertices at the bottom. Drag either red vertex vertically upward one unit at a time. The area of the square increases in this sequence: 10, 13, 18, 25, 34, 45, ... The rule is $n^2 + 9$.

Q10 Because the vertices of the square are on grid points, the corners of the outer dotted square must also be on grid points. Therefore, the area of the outer square is a whole number. Similarly, the outer triangles have whole-number sides and can be combined to form two rectangles that must also have a whole-number area. The area of the inner square is the difference between these two areas, so subtracting a whole-number area from another whole-number area must give a whole-number result for the area of the square.

Rather than subtracting, some students might prefer a dissection approach:

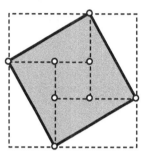

The dissection lines are all on the grid, so all the points marked are lattice points. The area of the shaded square is composed of the small inner square (which has whole-number sides) and the equivalent of two of the dotted rectangles (which also have whole-number sides). So, the shaded area is the sum of squares of whole numbers, which is also a whole number.

Exploring Algebra 1 with The Geometer's Sketchpad
© 2012 Key Curriculum Press

Squares and Square Roots

Dot paper is graph paper on which the grid is made up of dots instead of intersecting lines. Dot paper is useful for exploring math and creating art.

In this activity you'll use Sketchpad dot paper to explore several interesting properties of squares and square roots.

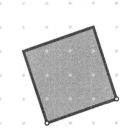

INVESTIGATE

1. Open **Square Roots.gsp.** You'll see a square constructed on dot paper. Drag one of the red corner points to change the size and orientation of the square.

Press and hold the **Custom** tool icon to choose one of these tools. Use the **Length** tool by clicking a segment. Use the **Area** tool by clicking the polygon interior.

2. Use the **Length** custom tool to measure the length of a side of the square, and use the **Area** custom tool to measure the area of the square. Drag one of the points and watch the measurements change.

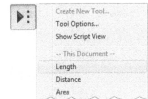

3. Calculate the square root of the area. To do this, choose **Number | Calculate** to open Sketchpad's Calculator. Choose **sqrt** from the Functions pop-up menu, click the area measurement in the sketch, and click **OK.**

Q1 What do you notice about the value of this calculation? Drag the points to make sure your observation is true for different squares. Complete these two statements:

The area of a square is the _____ of its side length.

The side length of a square is the _____ of its area.

Q2 Given what you know about squares, why do the relationships in Q1 make sense?

Q3 Use your sketch to find the square roots of 12 whole numbers less than or equal to 20. Round all decimals to two places.

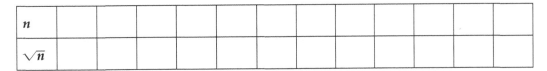

n												
$\sqrt{n}$												

Q4 Do you think it's possible to find the square root of any desired whole number using the method from Q3? Explain your reasoning.

PATTERNS

You can find many interesting number patterns in this sketch.

Q5 Set up the square so that it's perfectly balanced on its tip—in other words, so its diagonals are horizontal and vertical. The smallest such square has an area of 2. The next has an area of 8. You can think of this as a number sequence: 2, 8, ... Use your sketch to determine the first eight terms of this sequence. Write them down. Then describe the numeric rule for the sequence.

Pythagorean triples are numbers that satisfy the Pythagorean theorem, $a^2 + b^2 = c^2$.

Q6 When the square's base is horizontal, its side length is a whole number. But when the square is "slanted," its side length is usually an irrational number. In fact, you'll find only six slanted squares whose side lengths are whole numbers and that have at least one side completely within the sketch window. What are the side lengths of three such squares? (*Hint:* Think about Pythagorean triples.)

EXPLORE MORE

Q7 In Q5, you looked at a number pattern in the areas of squares as a vertex point is dragged. Draw the first four squares on grid paper or graph paper. Show how the area of the first square can be rearranged into 2 square units, the second square into 8 square units, and so on.

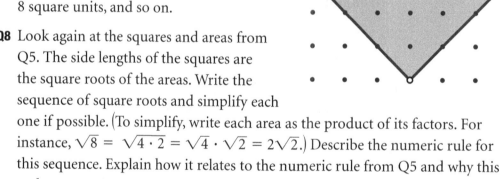

Q8 Look again at the squares and areas from Q5. The side lengths of the squares are the square roots of the areas. Write the sequence of square roots and simplify each one if possible. (To simplify, write each area as the product of its factors. For instance, $\sqrt{8} = \sqrt{4 \cdot 2} = \sqrt{4} \cdot \sqrt{2} = 2\sqrt{2}$.) Describe the numeric rule for this sequence. Explain how it relates to the numeric rule from Q5 and why this makes sense.

Q9 Find another number pattern as you drag a vertex in a systematic way, and write a numeric rule for that pattern.

Q10 You may have noticed that even slanted squares have whole-number areas. Write an argument explaining why. The figure at right may provide a hint.

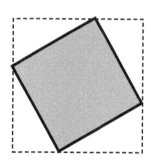

Exploring Algebra 1 with The Geometer's Sketchpad
© 2012 Key Curriculum Press

4

Solving Equations and Inequalities

Approximating Solutions to Equations

Students plot points on a number line to represent the left and right sides of an equation. They then drag a point to change the value of x continuously and approximate the solution to the equation.

Undoing Operations

Students model algebraic operations using "algebar" tools, use inverse operations to undo the original operations, and observe the symmetry of the resulting pattern.

Solving Linear Equations by Balancing

Students manipulate a balance model of equations, learn rules for making an equation simpler while keeping the two sides of the equation equal, and solve equations. The model works with both positive and negative weights.

Solving Linear Equations by Undoing

Students use a visual model of algebraic expressions to undo the expression on one side of an equation and find the solution to the equation.

Solving Linear Equations by Jumping

Students examine the motion of two rabbits who move at constant rates, and use distances and rates to write and solve equations of the form $a + bx = c + dx$.

Properties of Inequality

Students investigate arithmetic properties of inequality using a model in which the same operation is applied to two quantities that are initially unequal. Students observe that the direction of the inequality sometimes changes and formulate a rule to describe when the sign changes.

Solving Inequalities by Substitution

Students substitute a range of values into an inequality and determine the set of values that satisfies the inequality.

Solving Inequalities by Balancing

Students manipulate a balance model (actually an imbalance model). To solve inequalities, they use rules that won't change the state of the balance.

Solving Compound Inequalities

Students solve compound inequalities in one variable and plot the solution set on a number line.

Approximating Solutions to Equations

 ACTIVITY NOTES

This activity relates to the Dynagraph activities, in which independent and dependent variables are graphed on parallel number lines.

SKETCH

Q1 If the construction is correct, the unit point B on the upper number line should control the scale of both lines. Confirm that this is working before moving on.

Q2 The number line value of the new point should match the calculation on the left, $3x - 5$.

Q3 When the points *left* and *right* come together, the left and right sides of the equation are equal for that value of x. This happens when $x = 5$, and that is the solution to the equation.

Some students may make the mistake of saying that 10 is the solution, because that is what appears on both sides. Check for understanding.

OTHER EQUATIONS

Q4 a. $x = 16.6$ b. $x \approx -2.67$

The precision of students' answers will depend on the scale they choose.

Q5 Answers will vary. An equation like $x = x + 1$ works.

Q6 Answers will vary. An equation like $2x = x + x$ works.

Q7 a. $x = -5$ or 2 b. $x \approx 6.96$

The equation in part a is a quadratic equation with two solutions. If this is something new for students, ask them to explain why there are two answers. For the equation in part b, ask why the left-side calculation is undefined when x is dragged to the negative end of the number line.

VARIATIONS

You may wish to challenge students with other equations. Even if the solution is near zero, the expressions on each side of the equation may have very large magnitudes, making this model unwieldy. It is not actually necessary to see points *left* and *right* since you can see the calculations they represent. You can make this easier by subtracting one calculation from the other and finding the value of x for which the difference is zero.

 ACTIVITY NOTES

WHOLE-CLASS PRESENTATION

In this presentation students will observe how dragging a point on a number line can quickly generate many values to substitute in a simple equation, making it easy to find an approximate solution to the equation.

There are different ways to solve an algebraic equation in one variable: by algebraic manipulation, by graphing each side of the equation as a function and finding the point of intersection, and by substituting different values for the unknown to get closer to the answer. In this activity you will be demonstrating the last of these, by dragging a point along a number line to change the value of x continuously. This makes it easy to find approximate solutions quickly, even for equations that are quite difficult or impossible to solve analytically.

You will use the first page of the presentation sketch to solve the equation $3x - 5 = 2x$. The values of the expressions on the left and right sides of the equation correspond to two points labeled *left* and *right*. As you move x along the top number line, points *left* and *right* move accordingly along the bottom number line. Students look for the value of x that makes *left* and *right* coincide.

To present this activity to the entire class, follow the Presenter Notes and use the sketch **Approximating Solutions Present.gsp**.

Approximating Solutions to Equations

 Presenter Notes

Use these steps and questions to present this topic to the class.

1. Open **Approximating Solutions Present.gsp** and use it to introduce the presentation.

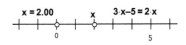

2. Go to page 2. Press the *Show Equation* button.

Define the term *substitute* if students don't already know it.

3. Explain that solving an equation means finding the value of *x* that makes the two sides of the equation equal.

4. Press the *Show x* button. Drag point *x* left and right to show how the value changes. Emphasize that *x* is a variable.

5. Press the *Show Left Side* button to show a point labeled *left*, corresponding to the value of the left side of the equation.

6. Drag *x* again so students can see how changing *x* changes the position and value of point *left*.

Have several students answer the questions in their own words.

Q1 Ask students how dragging *x* relates to trying different numbers for *x*. (One of their observations should be that dragging makes it easy to try a lot of numbers for *x* very quickly.)

Q2 Leave *x* someplace other than 2, and ask students if they can tell you where point *left* will go if you move point *x* to 2. (Answer: $3x - 5 = 3 \cdot (2) - 5 = 1$.)

7. Press the *Show Right Side* button, and drag *x* again. Avoid emphasizing the solution for the time being, and leave *x* at a value for which the equation is false.

Consider calculating the difference between the left- and right-side values. Students should predict that *left − right* = 0 at the solution.

Q3 Ask students whether the equation is now true or false. Then ask what they would expect to see if the equation were true. (Students should expect two things: The values will be equal, and points *left* and *right* will coincide.)

8. Drag *x* left and right. Ask whether each direction moves you closer to an answer or farther away. Ask if *left* will always be to the left of *right*. Drag *x* again until the points coincide.

Q4 Have students verify that this value of *x* does indeed give an approximate solution to the equation.

After changing the equation, you may need to change the scale to see the solution.

Use the remaining pages of the sketch to try different equations. Alternatively, double-click the existing calculations on page 2 to change the left and right expressions to solve any equation.

Before you try to find an exact solution to a problem, you may find it helpful to first approximate a solution. In real life, approximations may be good enough. For instance, if you are driving to Yellowstone National Park, you may be glad to know that you will be there in approximately $3\frac{1}{2}$ hours. It may not be possible to know *exactly* how long it will take you, but knowing the approximate time will help you plan.

SKETCH

Begin by creating two number lines. You will do this by creating two sets of coordinate axes, and then hiding the *y*-axis.

1. In a new sketch, choose **Graph | Define Coordinate System.** Then choose **Graph | Hide Grid.**

Click a point with the **Text** tool to show its label, or select the point and choose **Display | Show Label.**

2. There are two points shown on the screen, one at the origin and another one defining the unit distance on the *x*-axis. Show the point labels. The labels should be *A* and *B*. Select both points and choose **Measure | Distance.**

3. Construct a point *C* on the negative *y*-axis. Hide the *y*-axis.

4. To create the second number line, select point *C* and distance measurement *AB*. Choose **Graph | Define Unit Distance.** A message will appear warning you that you are creating a second coordinate system. Click **Yes.** Hide the new grid and the new *y*-axis.

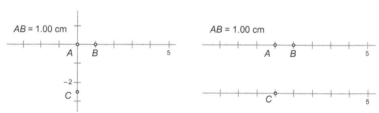

Q1 You should now have two number lines. What happens when you drag the unit point *B* on the upper number line?

To change a label, double-click it with the **Text** tool.

5. Construct a new point on the upper number line. Select the point and choose **Measure | Abscissa (x).** Change the measurement label to *x*, and change the label of the new point to *x* also.

The point and measurement correspond to the variable *x*. You will use them to build both sides of this equation:

$$3x - 5 = 2x$$

Exploring Algebra 1 with The Geometer's Sketchpad
© 2012 Key Curriculum Press

When you need to enter *x* into the expression, click the measurement *x* in the sketch.

6. Choose **Number | Calculate.** Compute the value of the expression $3x - 5$.

7. With the calculation selected, choose **Graph | Plot Value on Axis.** Click **Plot** in the dialog box. A new point should appear on the lower axis. If the point is not visible, drag point *x* across the screen until you see the new point.

8. Label the new point *left,* and use the **Segment** tool to connect it to point *x*.

Q2 What does the number line value of the new point represent?

The labels *left* and *right* refer to the two sides of the equation, not to the positions of the points on the number line.

9. Use the procedure from steps 6–8 (but using the calculation $2x$) to create another point on the lower line. This one represents the right side of the equation. Label it *right.*

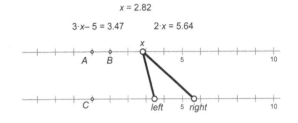

Q3 Drag point *x* until you see points *left* and *right* come together. When the points meet, what is the value of *x*? What is the solution to the equation?

OTHER EQUATIONS

You can edit the calculations to solve other equations. Double-click a calculation with the **Arrow** tool to get back to the Calculator window. After changing a calculation, you may need to adjust the scale to see the corresponding point.

Q4 Solve these equations.

  a. $31 - 3x = 2(x - 26)$

  b. $\dfrac{15(x - 3)}{4} = -40 - 7x$

Q5 Change the equations so that no matter where you drag *x*, there is no solution. Write down your equations.

Q6 Change the equations so that no matter where you drag *x*, *left* and *right* are always equal. Write down your equations.

Q7 So far, all of the equations have been linear, but this approximation method can work with any equation with one variable. Approximate the solutions to these.

  a. $x^2 + 5x + 3 = 2x + 13$

  b. $3\sqrt{x} = 2x - 6$

Undoing Operations

This activity is useful preparation for the activity Solving Linear Equations by Undoing. It makes the concept of undoing operations concrete and visual, and it provides a good introduction to the use of the algebar tools.

INTRODUCING THE ALGEBAR TOOLS

Q1 When you drag x, the tip of the $(x + 1)$ bar stays one unit to the right of the tip of the x bar.

Q2 Each tool creates a new bar showing the value of $a - b$. One tool creates a caption with parentheses, and the other creates a caption without them.

Q3 The **Indicator** tool shows a vertical line through any point so you can compare values easily. The **Measure** tool shows the numeric value of an algebar.

Q4 If you substitute 2 for x, the result is 6. Students should show work.

Q5 You must drag x to 0.5 to make $2(x + 1) = 3$.

Q6 To make the x bar and the $2(x + 1)$ bar the same length, x must be approximately -2. This represents the solution to the equation $x = 2(x + 1)$.

Q7 For $2(x + 1) = 0$, you must drag x to -1. For $2(x + 1) = 7$, you must drag x to 2.5.

Q8 The new bar, labeled $\frac{2(x + 1)}{2}$, matches the $(x + 1)$ bar.

Q9 The bottom bars are a reflected image of the top bars, with the middle bar serving as the mirror.

Q10 When $x = \frac{3x - 2}{4}$, the value of x is -2.

Q11 When x is exactly 2, the value of $\frac{3x - 2}{4}$ is 1.

Q12 The value of x is 6. The equation is $\frac{3x - 2}{4} = 4$

Q13 a.

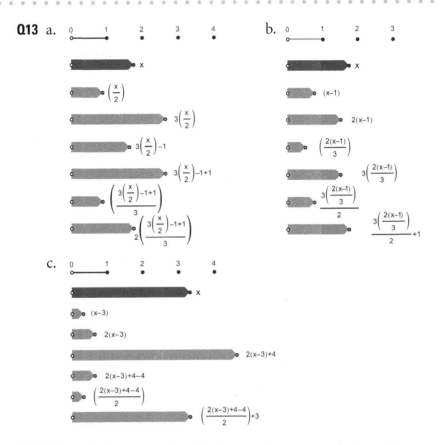

b.

c.

Q14 If you make *x* negative, the bars no longer match. Squaring a number cannot be undone uniquely. Of the two possible results for undoing this operation, the **sqrt(a)** tool produces only the positive one.

WHOLE-CLASS PRESENTATION

In this presentation students view a model of algebraic variables and expressions in which each value is represented by the length of a bar on the screen. Students see how dragging *x* makes it easy to substitute values in the expression and to find trial-and-error solutions to equations. They also see how undoing algebraic operations results in a distinctive visual pattern.

The mechanics of using the algebars tools to build the expressions are important for a hands-on student activity, but not for a whole-class presentation. To avoid these mechanics, the presentation sketch (**Undoing Operations Present.gsp**) skips the introduction of the tools and proceeds immediately to the process of creating and undoing constructions, providing buttons to avoid construction steps.

Most of the questions in this activity involve dragging *x*. As you present, drag *x* frequently to emphasize to students what it really means for *x* to be a *variable*.

Undoing Operations

Most algebraic operations have *inverses*—operations that undo the effect of the given operation. For instance, if you add 7 to a number, you can get your original number back by subtracting 7. In this activity you will construct *algebars* that model algebraic expressions and then construct additional bars to undo the algebra and get back to the original value.

INTRODUCING THE ALGEBAR TOOLS

As you work, drag *x* occasionally to see how the bars behave as *x* changes.

1. Open **Undoing Operations.gsp.** This sketch shows a bar representing the value of x and has tools to create algebraic expressions using x.

2. Drag the red point at the tip of the x bar left and right. Observe how it behaves.

3. Just below the x bar, create a new bar for the expression $(x + 1)$: Press and hold the **Custom** tool icon, and choose the **(a+b)** tool from the menu that appears. Then click the tool on five objects: the first white point below the red bar, the red point at the tip of the x bar, the caption "x" that labels the bar, the blue point beneath the number 1, and the number 1 itself.

Q1 Choose the **Arrow** tool and drag x back and forth again. How does the $(x + 1)$ bar behave when you drag x?

Every tool with both a and b in its name requires five clicks: one for the starting point, two for the value and caption of a, and two for the value and caption of b. For values, you can use either the point at the tip of an existing bar or one of the blue points representing constants.

4. Try out several of the tools to get a feel for how they work. Create the algebars shown below, and then make more bars of your own.

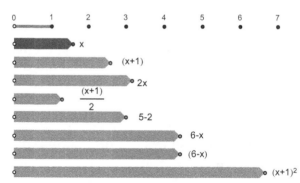

Q2 What's the difference between the **a–b** tool and the **(a–b)** tool?

Q3 Click the **Indicator** tool on the red point at the tip of the x bar. Then click the **Measure** tool on the red point and on the caption. What happens?

CREATING AND UNDOING AN EXPRESSION

Now you'll use these tools to model the expression $2(x + 1)$, and then you'll undo the result to get back to x.

5. Go to page 2. Use the **(a+b)** tool and then the **ab** tool to build the expression $2(x + 1)$. Put an indicator through x, and another through $2(x + 1)$.

Q4 If you substitute 2 for x, what should be the value of $2(x + 1)$? First find the answer by dragging x so its value is 2. Then check by substituting 2 into the expression and evaluating it on paper. Show your work.

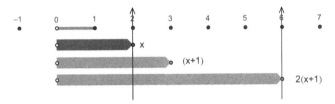

Q5 Adjust x so that $2(x + 1) = 3$. What value of x makes this happen?

Q6 Adjust x so that the x algebar and the $2(x + 1)$ algebar are the same size. What value of x makes this happen? What equation does this represent?

Q7 Use the algebars to solve these two equations: $2(x + 1) = 0$ and $2(x + 1) = 7$.

To get $2(x + 1)$, you started with x and performed two algebraic operations. Now you will perform the opposite operations to get back to the original value of x.

6. Use division to undo the multiplication operation: Divide the $2(x + 1)$ algebar by 2 by using the **a/b** tool.

Q8 What other algebar does this new bar match?

For step 7 you can either subtract 1 or add −1.

7. Use subtraction to undo the addition step.

Q9 Drag x back and forth and observe the pattern made by your five algebars. Describe the shape or symmetry that they show.

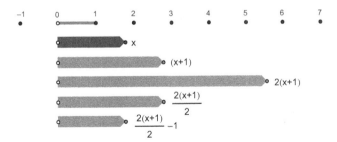

ANOTHER EXPRESSION

8. Use page 3 to build this expression:
$$\frac{3x - 2}{4}$$

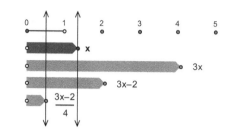

It took three operations to build the expression, so it takes three operations to undo it.

9. Add three more algebars to undo the expression and get back to the original value of x.

10. Put indicators on the points at the tips of the algebars for x and for $\frac{3x - 2}{4}$.

Q10 Drag x so that its length matches the $\frac{3x - 2}{4}$ algebar. What is the value of x, and what equation does this arrangement represent?

Q11 Drag x until it's exactly at 2. At what value is the $\frac{3x - 2}{4}$ bar?

Q12 Drag x to make the value of $\frac{3x - 2}{4}$ as close to 4.0 as you can make it. What is the value of x? What equation did you just solve?

EXPLORE MORE

Q13 Using the remaining pages, build each of the following expressions and then undo your operations to get back to the original value of x. Be sure to drag x to test your results.

a. $3\left(\frac{x}{2}\right) - 1$

b. $\frac{2(x - 1)}{3}$

c. $2(x - 3) + 4$

Q14 Build the expression x^2 and then use the **sqrt(a)** tool to undo it. What difficulty do you encounter? Describe the problem you see when you drag x, and explain why the problem occurs.

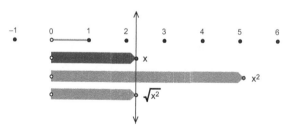

Exploring Algebra 1 with The Geometer's Sketchpad
© 2012 Key Curriculum Press

Solving Linear Equations by Balancing

 **ACTIVITY NOTES**

The balance used in this activity was made using the tools described on the Algebalance page of the presentation sketch. You can use these tools to create your own balance sketches.

EXPLORE THE BALANCE

Q1 The positive objects (1, 5, and x) weigh the pan down, and the negative ones (-1, -5, and $-x$) pull it up. (The behavior of x and $-x$ depends on whether x is positive or negative. This is explored later in the activity, so there's no need to worry about it yet.)

Q2 Answers will vary.

Q3 Answers will vary. The equation should match the list of objects from Q2.

Q4 Students should press *Setup Q4* before each part so that they have enough objects to perform the operation.

   a.   The pans remain balanced.

   b.   The right pan goes up and the left pan goes down.

   c.   The pans remain balanced.

   d.   The left pan goes down and the right pan goes up.

   e.   The pans remain balanced.

   f.   The pans remain balanced.

   g.   The left pan goes up and the right pan goes down (if students have not found and adjusted the x slider).

   h.   The pans remain balanced.

Q5 The rules are described later in the activity. Encourage students to answer this question in their own words.

Rule 1: You can drag the same kind of object from storage to each pan. (This rule is illustrated by Q4 parts a and e. Some students may also list f, which involves removing the same object from both pans.)

Rule 2: If you have matching positive and negative objects on a pan, you can remove them to the storage area. (This rule is illustrated by Q4 parts c and h.)

 ACTIVITY NOTES

SOLVE AN EQUATION

Q6 The balance on page 2 represents $2x - 1 = x + 5$.

Q7 Rule 1 allows you to drag identical objects to each pan.

Q8 Rule 2 allows you to eliminate pairs like x and $-x$. Students can use this rule twice, once on each pan.

Q9 The resulting equation is $x + (-1) = 5$. The sketch is limited in how it can display equations, and the appearance in the sketch is $x + -1 = 5$.

Q10 Students can remove a single combination of 1 and -1 from the left pan.

Q11 The resulting value of x is 6.

MORE EQUATIONS

Q12 Here are the steps, though students may change the order in some ways. Students may also combine similar steps such as b and c, or they may use step g to add three 1's to each pan.

   a. Add $-x$ to both pans: $2x + (-3) = 1x + (-1)$

   b. Remove x and $-x$ from the left: same equation

   c. Remove x and $-x$ from the right: same equation

   d. Add $-x$ to both pans: $x + (-3) = -1$

   e. Remove x and $-x$ from the left: same equation

   f. Remove x and $-x$ from the right: same equation

   g. Add 1 to both pans: $x + (-2) = 0$

   h. Remove 1 and -1 from the left: same equation

   i. Remove 1 and -1 from the right: same equation

   j. Add 1 to both pans: $x + (-1) = 1$

   k. Remove 1 and -1 from the left: same equation

   l. Remove 1 and -1 from the right: same equation

   m. Add 1 to both pans: $x = 2$

   n. Remove 1 and -1 from the left: same equation

   o. Remove 1 and -1 from the right: same equation

The last formula ($x = 2$) shows the value x must have to make the pans balance.

Q13 On page 4, the balance shows $2x + 5 = x + 2$.

Q14 To solve this equation, here are the steps:

   a. Add $-x$ to both sides: $x + 5 = 2$

   b. Add -1 to both sides: $x + 4 = 1$

   c. Add -1 to both sides: $x + 3 = 0$

   d. Add -1 to both sides: $x + 2 = -1$

   e. Add -1 to both sides: $x + 1 = -2$

   f. Add -1 to both sides: $x = -3$

Students may consolidate some of steps b–f by dragging more than a single -1 at a time.

Q15 A single x on the left pan pulls it up, even though it weighed it down in Q1. In Q1 (on page 1) x was positive, causing it to weigh the balance down. Here, x is negative, so it pulls the pan up. If you drag the x slider so that x is positive, this page will give the same result as page 1. If possible, have students try this by putting a single x on the left pan and then dragging the x slider to both positive and negative values.

Q16 The solution to $4x - 2 - x = x + 3 + x$ is $x = 5$.

EXPLORE MORE

Q17 When each pan has two identical columns, removing one column from each pan will keep the pans in balance. This is equivalent to dividing the number of objects on each pan by two.

Q18 a. $x = 3$ b. $x = -3$ c. $x = -4$

Q19 $-2x + 1 = -3x + 4$: $x = 3$

   $2x - 2 = 3x + 1$: $x = -3$

   $-x + 3 = -2x + 1$: $x = -2$

Q20 Students will make different equations. If possible, have some students show their equations and the method to use in solving them.

Q21 It may help students to study page 7 to see how the buttons on that page work.

CLASS DISCUSSION

Some fundamental principles arise during this activity.

Rule 1 is the addition property of equality. Have students explain this rule in their own words during the discussion. If they are already familiar with

this property by name, encourage them to make the connection between the concrete action of dragging objects on the screen and the abstract formulation of the property. If they are not familiar with the property by name, this may be a good opportunity to have students come up with their own name for the property that is more descriptive.

Different students will justify Rule 2 differently. A class discussion will help them develop insights into the concepts of additive identity and the additive inverses, and how they can be used to solve equations. There's no need to name these concepts; the important point is for students to realize that the sum of a number and its additive inverse is zero, so that removing the combination from one side of an equation leaves the two sides equal.

A discussion of the balance on page 4, and of Q15 in particular, provides an opportunity for students to recognize that $-x$ is not necessarily negative, any more than x is necessarily positive. This important realization can prevent many errors and misunderstandings later.

Finally, students are told several times during the activity to try to get a single x on a pan all by itself. The activity doesn't try to explain this strategy, so a class discussion is a good opportunity for students to think about this strategy and understand why it's useful.

WHOLE-CLASS PRESENTATION

In this presentation students see the effects of placing both positive and negative objects on the left and right pans of a balance, and connect the visible objects with a simplified formula corresponding to the current state of the balance. This presentation helps students develop a mental model of equations and of the techniques used to solve them.

Use **Solve by Balancing Present.gsp** and the Presenter Notes to conduct a presentation for the entire class.

Introduce the presentation.

1. Open **Solve by Balancing Present.gsp** and show how the balance works. You can press the numbered action buttons to show how the balance responds to the objects, or you can just drag objects to and from the balance.

   As you press the buttons or drag the objects, explain how an equation is like a two-pan balance, with equality corresponding to the pans being in balance and inequality corresponding to one pan being heavier than another.

Illustrate two rules students can use to solve equations.

2. Use page 2 to demonstrate one of the rules that makes it possible to solve equations by balancing: the addition property of equality. Press the *Show Premises* button to show the premises and to reveal buttons that allow you to illustrate the rule.

3. Use page 3 to demonstrate a second rule: Opposites on the same side of the equation add up to 0, and you can remove them.

Use the rules to solve several equations.

4. Use page 4 to show that you can solve simple equations using these rules. Press each numbered button in turn to show the various steps in solving the equation. Discuss how the solution becomes obvious once the balance has been manipulated so that there's only a single *x* left, on a pan by itself.

5. Use page 5 to solve another simple equation. The value of *x* on this page is negative, so that an *x* object pulls the balance up instead of weighing it down, and a −*x* object pulls it up. After you finish solving the equation, press the *Reset* button to clear the pans, and show what happens when you drag the *x* and −*x* objects on and off the pans. Discuss this with your students, and encourage them to describe and explain this behavior in their own words.

If students are not ready to think of the operation on this page as multiplying by $\frac{1}{2}$, describe it as dividing by 2.

6. Use page 6 to suggest a third rule (the multiplication property of equality). In this example both pans have two identical columns of objects. By removing half of what's on each pan (removing half the weight on each pan), the pans remain balanced. Similarly, doubling the number of objects on each pan keeps them balanced.

7. Page 7 shows the solution of a more complicated equation using the balance.

8. Page 8 is blank. You can use it to set up and solve any of the problems from the student activity pages.

Finish with a class discussion. As part of the discussion, encourage students to ask questions about and to justify these techniques for isolating *x* on one side of the equation while keeping it balanced.

Solving Linear Equations by Balancing

To solve a complicated equation, you can make it simpler while keeping the two sides of the equation equal. In this activity you will use a Sketchpad balance to show an equation and solve it by using operations that keep the sides balanced.

EXPLORE THE BALANCE

1. Open **Solve by Balancing.gsp.** Experiment by dragging objects from the storage area (on the left of the dividing bar) to the left or right balance pan.

Q1 Which objects weigh a pan down, and which ones pull it up?

Q2 Find a combination of weights and balloons different from the one shown below that balances the two pans. List the objects you put on each pan.

Q3 Write down the algebraic equation that corresponds to your arrangement of weights and balloons. Press the *Show Formula* button to check your answer.

When the pans are balanced, some operations disturb the balance and others do not.

Q4 Try each of these operations and write down how it affects the balance. Before each operation, press *Setup Q4* to make sure the pans are balanced and that each pan contains enough items to carry out that operation.

In your answer, state whether the operation makes the left pan heavier, makes the right pan heavier, or leaves the two pans in balance.

   a. Drag a 1 from the storage area onto each pan.

   b. Drag a 1 from the right pan to the left pan.

   c. Drag a 1 and a -1 together from the left pan to the right pan.

   d. Drag a 1 from the storage area onto the left pan and a -1 onto the right pan.

   e. Drag a -5 onto each pan.

   f. Remove an x from each pan by dragging to the storage area.

   g. From the storage area drag an x onto the left pan and a 5 onto the right pan.

   h. Drag an x and a $-x$ from the right pan to the storage area.

Q5 Write down at least two rules to describe things you can do that will keep the pans balanced. For each rule, write down which parts of Q4 illustrate the rule.

SOLVE AN EQUATION

Q6 Go to page 2. What equation does this balance represent? Press the *Show Formula* button to check your answer.

In the next few steps you will use these balancing rules:

Rule 1: You can drag the same kind of object from the storage area to each of the pans. (For example, you can drag a 5 to the left pan and a 5 to the right pan.)

Rule 2: If you have matching positive and negative objects on the same pan, you can remove them to the storage area. (For instance, if the left pan has both a 5 and a −5, you can remove them both.)

Make sure the pans stay in balance after each operation.

Q7 Drag a −x from the storage area onto each pan. Which rule allows you to do this?

Q8 Any time you find both an x and a −x on the same pan, you can remove them to the storage area. Which rule is this? How many such combinations can you find? Remove them now.

Q9 What is the resulting equation?

Q10 Drag a 1 from the storage area onto each pan. Any time you find both a 1 and a −1 on the same pan, you can remove them to the storage area. How many such combinations can you find and remove?

Q11 What is the resulting value of x? Press the *Show x* button to check your result.

MORE EQUATIONS

You have solved the equation when you have x by itself on one pan and only numbers on the other pan.

2. On page 3, use the rules about moving objects to eliminate as many objects as you can and to leave the last x all alone on its pan.

Q12 Write down each step that you follow, and write down the equation for the balance after each step.

Q13 On page 4, what equation does the balance show?

Q14 Use the two rules to add and remove objects until you get x on a pan by itself, with only numbers on the other pan. Each time you use Rule 1, write down what you did and the resulting equation. What is x?

Q15 Remove all the objects from the pans by pressing the *Reset* button, and then move a single x onto the left pan. Does this object weigh the pan down or pull it up? Is your answer the same as it was for Q1? If not, why not?

Q16 Go to page 5 and add objects to the pans to model the equation $4x - 2 - x = x + 3 + x$. Then solve the equation by following the two rules.

EXPLORE MORE

Q17 There's one more rule you can use with the algebra balance. Go to page 6 and notice that each pan has two identical piles of objects. Remove one pile from the left pan and one pile from the right pan. Do the pans stay in balance? What mathematical operation did you perform on each pan?

Q18 With the help of this third rule, solve these three equations:

    a. $2x - 1 = 5$ b. $4x + 3 = 2x - 3$ c. $4x + 5 = x - 7$

Q19 Page 7 contains several buttons to create different arrangements of objects. For each button, press the button and write down the equation that results. Then use the various rules you've learned to rearrange the objects and solve the equation, and write down the solution that you find.

Q20 Page 8 is a blank page. Show the value of x, adjust the slider, and arrange objects to make a balanced equation of your own choosing. Then hide the x value and slider, and challenge a friend to solve your equation.

Use Sketchpad's Help menu to find out how to make and use movement buttons.

Q21 Page 9 is also a blank page. Make a movement button to move objects out of the storage area and onto the pans to make an equation. Then make a movement button for each step in solving the equation. Use the movement buttons to demonstrate your problem to your classmates.

Solving Linear Equations by Undoing

The reflection symmetry shown by the algebars in this activity is helpful to students conceptually; it gives a visual sense of the way in which undoing operations returns you to your starting place.

SOLVE AN EQUATION

Q1 The indicator allows you to tell when the value of the expression is close to 2. When the indicator is close to 2, the value of x is within a few hundredths of 5.00. (If students have not changed the scale, they cannot drag x to exactly 5.00. This is intentional, to emphasize the limitations of the guess-and-check process.)

Q2 This method is called "guess-and-check" because dragging x to different positions is a way of guessing possible solutions and observing the position of the indicator is a way of checking how close you are to a solution.

Q3 When you drag x back and forth, the new bar doesn't change, but the $x + 3$ bar does. The new bar matches the $x + 3$ bar only when x has been dragged to make the value of the expression 2.

Q4 To perform the inverse of adding 3, you can either add -3 or subtract 3. Either method will work in the next step.

Q5 The sequence is $\frac{(x + 3)}{4} = 2$, $(x + 3) = 2 \cdot 4$, and $x = 2 \cdot 4 - 3$. This sequence matches the algebraic steps normally used to solve the equation.

Q6 The result of measuring is $\frac{(1 + 5)}{2} = 3$, so $x = 3$. Here are the algebars:

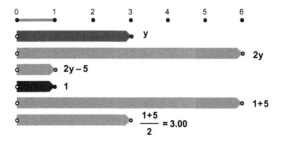

Q7 The solution to the equation $\frac{2(a+5)-1}{3} = 1$ is $a = -3$. As with the earlier problems, the algebars that undo the expression must be mirror images of the ones that build the expression. Here are the algebars:

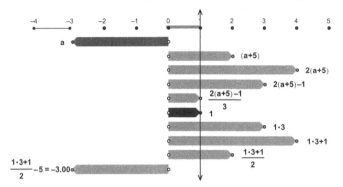

Q8 The solution is $n = 6$. Here are the algebars:

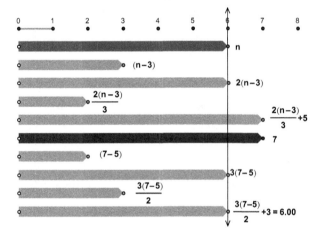

WHOLE-CLASS PRESENTATION

To present this activity to the whole class, use the Presenter Notes together with the sketch **Solve by Undoing Present.gsp.**

In this presentation students will observe a dynamic "algebar" model used to create an algebraic expression involving *x* and will see how performing inverse operations can undo the expression and find the original value of *x*.

1. Open **Solve by Undoing Present.gsp** and press the *Show Problem* button. Explain that to find *x*, the first step is to build the expression on the left.

2. Press the *Show x* button. Drag *x* left and right to show that it really is a variable. Tell students "We need to find the value of *x* that will make this expression equal to 2, so first we'll use *x* to build the expression."

3. Press *Add 3* and drag *x* to show how the algebar really does represent $x + 3$.

4. Press *Divide by 4* and drag *x* again. Ask students to stop you when *x* is at the right position. Encourage students to coax you to move *x* a bit to the left or right to make it closer.

5. Point out that it's hard to tell when you're in exactly the right spot, and press the *Show Indicator* button. Adjust *x* once more.

Q1 Ask students if the value of *x* is now exactly correct. (You have probably not been able to get closer than about 0.01 to the answer. More importantly, this is a guess-and-check solution method that may not yield an exact answer.)

6. Tell students that to get a precise answer, they must work backward from known facts, and it's a fact that the right side of the equation is 2. Press the *Show 2* button.

Q2 Ask, "To work back from the 2 to the *x*, the first thing we undo must be the last thing done. What operation was that?" (Students will answer, "Division by 4.")

7. Press the *Undo Division* button. Call students' attention to the fact that this new algebar, $2 \cdot 4$, is exactly the same size as the $x + 3$ bar. Write an equation.

Q3 Ask students what operation to undo next. (The answer is "Add 3.") Then ask them to give two different ways to undo the operation of adding 3. You should get responses of both "Subtract 3" and "Add -3."

Q4 Click the *Undo Addition* button and ask students which of the two methods was actually used. Write another equation.

Q5 Ask students what the solution is and whether they think this solution is exact or approximate. Develop this into a discussion.

Q6 Ask students whether the bars will always form such a precise pattern. Drag *x* to show that the pattern holds only when *x* is dragged nearly equal to the solution.

8. If there's time, use the remaining pages to explore some of the problems from the student activity.

Solving Linear Equations by Undoing

Algebra problems are often written in the form of equations that give you certain clues about a variable without stating its value directly. One method for finding the unknown value of the variable uses inverses to undo the steps that have been performed on the original variable.

SOLVE AN EQUATION

Algebars are bars that represent algebraic expressions. You can observe their behavior as the value of a variable changes.

1. Open **Solve by Undoing.gsp.** This sketch uses *algebars* to show the expression $\frac{x+3}{4}$. Drag x back and forth to observe the behavior of the different steps in the expression.

Your task is to find the solution to $\frac{x+3}{4} = 2$. You'll do it first by guess-and-check.

Press and hold the **Custom** tools icon and choose **Indicator** from the menu that appears. Click the tool on the point at the tip of the bar.

2. Use the **Indicator** custom tool to put an indicator through the point at the tip of the bottom green bar.

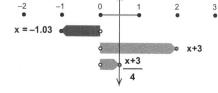

Q1 Drag x and observe the indicator. How does the indicator allow you to tell when you're close to the solution—that is, when $\frac{x+3}{4}$ is close to 2? Drag x until the indicator is as close to 2 as you can make it. What is the value of x?

Q2 Why is the method you just followed called "guess-and-check"?

Sometimes a guess-and-check solution is good enough. In this activity you will find exact solutions.

Now you'll use inverse operations to "undo" your way to an exact solution, starting with the final value of the expression and working your way back to the value of x.

3. According to the equation, the final value of the expression is 2. Use the **Value** custom tool to create an algebar with a length of 2. Click the tool on three objects: the first unused white point (where you want the bar to start), the blue point representing the value 2, and the caption 2 above the blue point.

The last step in creating $\frac{x+3}{4}$ was division by 4. Use the inverse of that operation (multiplication by 4) to undo the division.

Click the tool on five objects: the white point where you want to start, the blue point and caption for the 2 bar, and the blue point and caption for the number 4.

4. Use the **(a*b)** custom tool to multiply the 2 algebar by 4.

Q3 Is the resulting bar the same length as the $x + 3$ bar? Write this as an equation. Drag x back and forth. Does the bar stay the same length? Why or why not?

5. Use **Edit | Undo** to put x back where it was. (Leaving x at the approximate solution helps you to see the symmetry in the algebars.)

Q4 The first operation used to build the expression $x + 3$ was adding 3. Think about the inverse of this operation. How could you perform the inverse using addition? How could you perform the inverse using subtraction?

Exploring Algebra 1 with The Geometer's Sketchpad
© 2012 Key Curriculum Press

6. Pick one of these two methods. Choose the appropriate custom tool and use it to do the inverse operation to the 2 • 4 bar.

7. You have undone all the steps in the original expression, so this new bar shows the solution to the equation. Use the **Measure** tool to find its length.

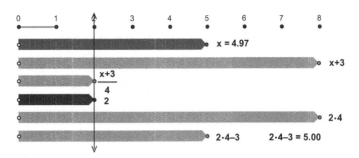

Q5 Write down the sequence of equations for the equivalent bars, starting with the two middle bars and moving outward to the top and bottom. To get you started, the equation for the two middle bars is $\frac{x+3}{4} = 2$.

SOLVE ANOTHER EQUATION

On page 2 are algebars that you will use to solve the equation $2y - 5 = 1$.

The pattern formed by your new bars should be the reverse of the pattern of the original bars for the expression.

8. First drag y so that the bottom bar has a value close to 1. This position of y will help you see the symmetry of the bars as you construct the inverses.

9. Use the **Value** tool to create a new bar with a length of 1.

10. Use the **a+b** tool to perform the inverse of the subtraction in the expression.

11. Use the **a/b** tool to perform the inverse of the multiplication.

Q6 Use the **Measure** tool to measure your last bar. What is the solution to the equation $2y - 5 = 1$?

EXPLORE MORE

Q7 On page 3 is a more complicated equation. Use the appropriate tools to undo the operations that were performed to build it. What is the solution?

Q8 On page 4, construct this expression: $\frac{2(n-3)}{3} + 5$. Then construct an algebar with value 7, and additional algebars to solve the equation $\frac{2(n-3)}{3} + 5 = 7$.

Solving Linear Equations by Jumping

RABBIT RACES

Give students time to experiment with the sketch and run some races. See what questions students have about the motion of the rabbits. Note that the jump size and starting positions take on integer values only.

Q1 Quincy has a lead of 12 units. Since Percy moves 4 units with each jump while Quincy moves only 2, this lead diminishes by 2 units per jump. The 12-unit lead divided by 2 units/jump equals 6 jumps.

Running the race in slow motion may allow students to see how the lead diminishes. For Q1, students might also reason using guess-and-check to determine the location of each rabbit after a certain number of jumps or repeatedly add the jump sizes to the initial position.

Q2 Since Percy moves faster in this race, it takes fewer jumps to catch up. Since Percy's jump size is now 5 units, Quincy's lead diminishes by 3 units per jump. The 12-unit lead divided by 3 units/jump equals 4 jumps.

Q3 At $x = 0$, Percy is at -3 on the number line. His position increases by 4 with each jump.

Q4 After 10 jumps, Percy is at 37 on the number line. You can calculate this by adding 4 ten times to the initial position of -3. For any number of jumps, 4 times the number of jumps added to the initial position will give the current position.

Q5 The equation $p = -3 + 4x$ gives Percy's position, while $q = 9 + 2x$ gives Quincy's.

Q6 $-3 + 4x = 9 + 2x$

Q7 If students are not familiar with writing equations such as these, they can spend additional time on page 4 setting up races and making tables, and then checking to see if their equations are correct.

Q8 When $x = 6$ jumps, $p = -3 + 4x = 21$ and $q = 9 + 2x = 21$, so $x = 6$ must be a solution to the equation.

Q9 At the start they are separated by 12 units, which is the difference between c and a.

Q10 Decrease by 2, since 2 is the difference between b and d.

Q11 As in Q1, divide the 12-unit lead by 2 units per jump.

 ACTIVITY NOTES

Q12 Percy will catch Quincy since the lead is still 12, and it diminishes at a rate of 2 units per jump. Algebraically, the equation has the same solution because for $x = 6$, both expressions have a value of 27.

Q13 The rabbits both move to the left since the jump sizes are negative. The equation has the same solution, however, since Percy still has a lead of 12 and the lead still diminishes at a rate of 2 units per jump.

Q14 Here, Percy has the lead. Quincy will catch up by moving at a faster rate. The solution to the equation is still $x = 6$, as the lead is 18 at the start, and diminishes by 3 for each jump made, since 7 is 3 greater than 4.

Q15 The rabbits run toward each other, starting 25 units apart. The gap between them diminishes by 5 for each jump, since Percy moves 2 to the left, and Quincy 3 to the right, with each jump. The 25-unit difference divided by 5 units per jump equals 5 jumps.

EXPLORE MORE

Q16 If the jump sizes are the same and the starting positions are different, the distance between the rabbits remains constant. The value of b and d is the same.

Q17 In such a race, Quincy starts with a lead of 12. Because Quincy's jump size is 2 greater than Percy's, this gap does not diminish—it widens. The equation has a solution of $x = -6$. To show the solution, move the green jump value to the left of zero and run the race.

In this presentation you will model the motion of two rabbits who move at constant rates, and use distances and rates to write and solve linear equations.

1. Open **Solve by Jumping.gsp.**

2. Press the *Jump Once* button to demonstrate how the race will proceed with the given values of *a*, *b*, *c*, and *d*. Press the *Reset* button, change the values, and show the first two jumps in a different race to demonstrate how the races work.

Q1 Go to page 2 and ask, "How many jumps will it take for Percy to catch up to Quincy?" Follow up with "How did you make your prediction?"

Discuss the different ideas students have, which may include adding the jump size of each rabbit repeatedly to their initial positions until the positions are equal.

3. On page 3, run the race with the *Race!* button. Then press *Reset* and *Show Tables*. Use *Jump Once* to run the race one jump at a time, double-clicking the table after each jump. Ask what happens to the distance between the rabbits with each jump made.

Q2 Reset again and set Percy's jump size to 5. Ask, "Will Percy catch up to Quincy after fewer jumps or more jumps? Why?"

Ask students to make predictions for a variety of different races, each time asking them to explain how they arrived at their prediction. Focus the questions on how the initial distance between the rabbits and the difference in their jump sizes determines how many jumps it takes for Percy to catch up.

For students who are not familiar with writing expressions for linear sequences like $-3, 1, 7, 11$, spend time creating some different races and discussing the idea of how $-3 + 4 + 4 + 4 + 4 + 4$, or $-3 + 4(5)$, is the position of Percy after 5 jumps, and how $p = -3 + 4x$ can be used to determine the position after *x* jumps.

Once equations for the position of each rabbit are determined, allow students to note that $-3 + 4x = 9 + 2x$ is *not* a true statement for most values of *x*. Explain that solving the equation means finding a value of *x* for which the statement is true.

Use Q8–Q15 in the activity. Focus on interpreting the equations in terms of the races and on understanding why the solutions to the equations in Q12–Q14 are all the same.

Q3 Ask students to write another equation that has a solution of 6.

Q4 Ask students to think in terms of the race to explain why the equation $-3 + 2x = 9 + 4x$ does not have a solution of 6.

Exploring Algebra 1 with The Geometer's Sketchpad
© 2012 Key Curriculum Press

Solving Linear Equations by Jumping

In this activity you will use "rabbit races" as a way to think about and solve algebraic equations.

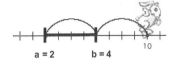

a = 2   b = 4   10

RABBIT RACES

1. Open **Solve by Jumping.gsp.**

Try to predict the outcome of the races in advance.

2. Press the *Race!* button to run a race. Try the other buttons to see what they do, and drag the numbers (not the bars) to change the starting positions and jump sizes. Run races with different starting positions and jump sizes.

Q1 On page 2, predict how many jumps it will take for Percy to catch up to Quincy. How did you make your prediction?

Drag the numbers themselves to change their values.

3. Check your prediction by running the race. Then press *Reset* and *Show x.* Drag the green number *x* to run the race again, this time in slow motion.

Q2 Reset again and set Percy's jump size to 5. Will Percy catch up to Quincy after fewer jumps or more jumps? Why?

WRITING EQUATIONS

The bottom row of the table is not fixed. The numbers in the bottom row change when the sketch changes.

4. On page 3, press the *Show Tables* button, and double-click each table to enter the current values permanently.

x	Percy		x	Quincy
0.00	-3.00		0.00	9.00
1.00	1.00		1.00	11.00
2.00	5.00		2.00	13.00

5. Run the race one jump at a time by clicking the *Jump Once* button. After each jump, double-click the tables again. Observe the values that appear in each of the tables.

Q3 What is Percy's position at the beginning of the race, when $x = 0$? By how much does his position change after each jump?

Q4 Where will Percy be after ten jumps? Without using the table, how can you quickly calculate Percy's position after any number of jumps?

Q5 Write your answer to Q4 as an equation that gives Percy's position for any number of jumps. Write another equation for Quincy's position. Use p and q to stand for the positions of the rabbits.

At the moment when Percy catches Quincy, the positions of the two rabbits are the same. You could write this in the form of an equation: $p = q$.

Q6 Write the equation $p = q$ in another form, using your expressions from Q5 in place of p and q.

SOLVING EQUATIONS

Solving an equation means finding the value of a variable that makes the left and right sides equal. In this section you'll find the value of x when Percy catches Quincy.

Q7 On page 4, press the *Show Equation* button. Does the equation match your answer to Q6?

Q8 Where will each rabbit be after six jumps? What does this answer indicate about the solution to $-3 + 4x = 9 + 2x$?

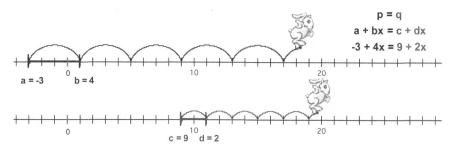

Q9 How far apart are the rabbits at the start of the race? How can you calculate this from a, b, c, and d in the equation?

Q10 By how much does the distance between the rabbits increase or decrease for each jump? How can you calculate this from a, b, c, and d in the equation?

Q11 How can you use your last two answers to predict when Percy catches Quincy?

Q12 Change the race so that the rabbits' positions are determined by $-3 + 5x = 9 + 3x$. Explain how you can predict that Percy will still catch Quincy in the same number of jumps. Why does the equation $-3 + 5x = 9 + 3x$ have the same solution as $-3 + 4x = 9 + 2x$?

Drag the top number line to move all three number lines left or right to see the outcome of the race.

Q13 Run this race: $9 - 5x = -3 - 3x$. Why do the rabbits both move to the left? Why does this equation still have the same solution as the one in Q12?

Q14 Set up a race to solve $21 + 4x = 3 + 7x$. What happens? How can you use your methods to solve it anyway, even though the rabbits run off the screen?

Q15 Run this race: $17 - 2x = -13 + 3x$. How is it different from the other races you have run? Explain how you can still use the same methods to solve it.

EXPLORE MORE

Q16 Imagine a race in which Percy and Quincy always stay the same distance apart. What would the two expressions for their positions look like?

Q17 Set up this race: $2x + 8 = 4x + 20$. What happens? Why? Does the equation have a solution? Try to figure out how to show the solution in your sketch.

Exploring Algebra 1 with The Geometer's Sketchpad
© 2012 Key Curriculum Press

Properties of Inequality

 ACTIVITY NOTES

EQUALITY RULES

Q1 Using only the buttons, it is not possible to make the heights of the bars different from each other. They are equal to begin with. For any operation, there is only one possible outcome, so they will remain equal.

INEQUALITY RULES

Q2 Addition and subtraction preserve the inequality. These operations translate the ends of both bars vertically. Whichever bar is higher at the beginning must remain higher.

Multiplying or dividing by a positive number can cause the values to be closer or further apart, but the greater value will always remain greater.

Q3 Multiplying or dividing by a negative will consistently change the direction of the inequality sign. The general rule is that whenever you multiply or divide by a negative number, you must change the direction of the sign.

There is one more case that students may notice. If you multiply by zero, the inequality becomes an equality: $0 = 0$. Division by zero is not possible.

SUGGESTED EXTENSION

Ask students what happens when they square both sides of an inequality. They cannot use the sketch for this but will have to reason it out by themselves. A general rule is more difficult in this case. The side that has the greatest *magnitude* will have the greatest value after squaring. Therefore, there can be no general rule for unknown values.

WHOLE-CLASS PRESENTATION

In this presentation students see a visual model in which the same operation is applied to two different quantities that are initially equal, and observe that the quantities remain equal. When the model is changed so that the initial quantities are unequal, students observe that the sign of the inequality sometimes changes. They formulate a rule to describe when the sign changes.

 ACTIVITY NOTES

To do the presentation, follow the directions on the student activity page. When investigating equality, be sure to try both positive and negative numbers for each operation. The easiest way to change one of the numbers from positive to negative may be to select the number and hold down the minus sign on the keyboard until the number changes sign. (You can press the **+** and **−** keys to increase or decrease the value of a selected parameter.)

In step 4, drag each marker in turn, both up and down, to verify that the inequality sign between the two bars is always accurate.

In Q2, when students first detect a change in the direction of the inequality, use **Edit | Undo** to revert to the previous step. Then perform the operation again. Students may find this result unexpected and want to observe the bars while the operation is performed several times. This is a good time for a class discussion in which students explain this behavior in their own words.

After students have formulated a general rule with which they are comfortable (that they must change the direction of the inequality when multiplying or dividing by a negative value), ask them what happens at the boundary between positive and negative. In other words, they haven't said what happens if the value being used is exactly zero. Ask them to make a conjecture first, before trying it in the sketch.

Similarly, ask them to make conjectures about what will happen if the two values are divided by zero, and then test the conjectures.

Properties of Inequality

A good general rule for equations is that if you do the same thing to both sides, they will always remain equal. This rule applies to addition, subtraction, multiplication, division, and any other operation. It would be very convenient if we could apply the same rule to inequalities. Before we jump to any conclusions, however, let's check to see how inequalities behave.

EQUALITY RULES

1. Open **Properties of Inequality.gsp.**

The two bars represent values on two sides of an equation. The numbers they represent are shown on the number line. On the left side are four buttons for the four elementary arithmetic operations.

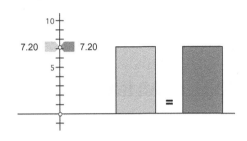

2. Press the *Add* button. This adds 2.6 to both sides of the equation.

To change the parameters next to the buttons, select a number and press the + or − key, or double-click the number to type a new value.

3. Edit the value of the number next to the *Add* button. Change it to −3.8. Press the button again. This time, you are adding −3.8 to both sides.

Q1 Experiment with all four arithmetic buttons. Try at least one positive and one negative number with each of them. Using only these buttons, is it possible to make the heights of the bars different from each other? Explain.

INEQUALITY RULES

4. On the number line, drag the red marker so that it is higher than the blue marker. Notice the change in the symbol between the two bars. Now the red value is greater than the blue value.

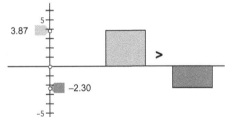

Q2 Once again, use all four arithmetic buttons with positive and negative numbers. Which operations can you use without changing the inequality sign?

Q3 Which operations change the inequality sign? What is a general rule to follow in these cases?

Solving Inequalities by Substitution

SUBSTITUTION ON THE NUMBER LINE

1. This inequality may be difficult to interpret at first. Each side is a calculation. In addition to the formula in each calculation, an equal sign and the numerical value also appear.

Q1 The red line segment appears if and only if x is a solution to the inequality.

Q2 $x > -4$

Q3 a. $x > 5$ b. $x < -1$

NONLINEAR INEQUALITIES

Q4 The main purpose of this section is to warn students that the computer may not be telling them the whole story. It can perform calculations for them, but it cannot think for them.

a. The range $x > 2$ is part of the solution, and students may see this by tracing it using the scale that is in use when the file is first opened. They will have to change the scale in order to see that the range $-12 < x < -10$ is also in the solution set.

b. If students trace the solution to this inequality using the original scale of the sketch, it will appear that the solution is $x > 0$. Have them change the scale so that the number line range on the screen is between -0.1 and $+0.1$. They will see that the solution is actually $-0.02 < x < -0.01$ or $x > 0.01$.

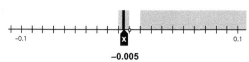

WHOLE-CLASS PRESENTATION

To present this activity to the whole class, use the sketch **Inequalities by Substitution Present.gsp.** This sketch follows the same steps as the paired/individual activity but has larger text and contains buttons to perform some of the functions.

Here are the differences:

Q1 You can use the *Animate x* button to animate *x* back and forth alone the number line.

2. You can use the *Toggle Segment Trace* button to turn tracing on and off.

Q3 You can use pages 2 and 3 to investigate these two inequalities, with no need to use the Calculator.

Q4 Use pages 4 and 5 to investigate these two inequalities.

Finish with a class discussion. Ask students to describe what they noticed about solving inequalities that is different from solving equations and what is the same. Discuss the results of Q4; these results should help convince them that it would be nice to have a more systematic way of finding solutions, so that they won't miss the less obvious aspects of the solution.

Solving Inequalities by Substitution

To solve an equation with one variable, you find the number(s) that can be substituted for the variable in that equation. If the substituted number makes the equation true, then it is a solution. You can solve inequalities the same way, but the solution is usually a continuous range of numbers, not just one or two.

SUBSTITUTION ON THE NUMBER LINE

A crude way to solve an inequality would be to substitute every possible number for x and see which ones satisfy the inequality. You can't do this for infinitely many real numbers. However, with a computer you can check many numbers quickly.

1. Open **Inequalities by Substitution.gsp.**

This sketch shows the inequality $4x > x - 12$. Marker x is on a number line, and its value is substituted into two calculations: $4x$ on the left and $x - 12$ on the right. Depending on the value of x, the inequality may be true or it may be false.

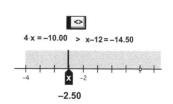

Q1 Drag marker x across the screen. For certain values of x, a red line segment appears above the point. What does the red line segment indicate?

To turn on tracing for the segment, select it and choose **Display | Trace Segment.**

2. Mark the interval on which x satisfies the inequality by turning on tracing for the segment and dragging the point along the number line.

Q2 What values of x satisfy the inequality $4x > x - 12$?

3. To edit a calculation, double-click it. To enter x in the Calculator, click the value of x in the sketch. To change the direction of the inequality, press the button above the inequality sign. To erase old traces, press the *Erase Traces* button.

Q3 Edit the calculations and find the solution sets to these inequalities:

   a. $2x + 13 < 6x - 7$ b. $5 - 3x < 4 - 4x$

NONLINEAR INEQUALITIES

Your computer helps you do some amazing things, but it can also limit you if you depend on it too heavily. Numbers may be too big to fit on the screen, and details can be too small to notice.

To adjust the scale, drag one of the tick mark labels below the number line.

Q4 Try to find solutions to these nonlinear inequalities, then adjust the number line scale and see if you missed anything.

   a. $240 - 20x^2 < x^3 + 76x$ b. $10000x^3 > x$

Solving Inequalities by Balancing

 ACTIVITY NOTES

The balance used in this activity was made using the **Algebalance.gsp** sketch. You can use these tools to create your own balance sketches; instructions accompany the sketch.

EXPLORE IMBALANCE

Q1 Rule 1: Dragging the same type of object onto each pan does not change the balance.

Q2 Rule 2: Removing an object and its opposite (for instance, x and $-x$, or 1 and -1) from a pan does not change the balance.

Q3 The pans represent the inequality $x + 3 > 2x - 1$.

Q4 The solution is $4 > x$.

Q5 The original inequality is $3x - 2 < 2x - 3$. To solve it, follow these steps:

    a. Move two $-x$ objects onto each pan. (Rule 1)

    b. Remove zeros, resulting in $x - 2 < -3$. (Rule 2)

    c. Move two 1 objects onto each pan. (Rule 1)

    d. Remove zeros, resulting in $x < -1$. (Rule 2)

Q6 Removing exactly half of the objects on each pan cannot change the balance. Though the example involves multiplying by one-half (or dividing by two), Rule 3 actually states that you can multiply or divide both pans by any positive number. (Negative numbers work here for equality, but not for inequality.) Students will formulate Rule 3 in different ways, some more general and some more limited. This is a good question to pursue in a class discussion.

Q7 The inequality is $x + 3 > 3x - 3$. The solution is $x < 3$.

CLASS DISCUSSION

Discuss with the class the differences between using the balance to solve equations and using it to solve inequalities.

A very important difference to note is that it's possible to violate the rules without any visible effect, depending on the nature of the violation and the value of x. For instance, if the correct solution is $x < 3$, and a student makes a mistake and ends up with $x < 4$, she will not see any change in the state of

the balance, even though she now has the wrong solution set. The state of the balance gives information for only a single value of *x*, not for the entire solution set.

A discussion of Rule 3 is also important, both because some students will formulate it in different ways in Q6 and because the rule is different for inequality than it is for equations. Students will understand Rule 3 better if they have done the activity Properties of Inequality.

Finally, review the value of the strategy in which students try to get a single *x* on a pan all by itself. Understanding the value of this arrangement of the balance will help students in solving equations and inequalities symbolically.

WHOLE-CLASS PRESENTATION

Use **Inequalities by Balancing.gsp** to conduct a presentation for the entire class. Follow the steps of the student activity, and involve the class in the process of answering the questions in the activity.

Solving Inequalities by Balancing

Just as you solve equations by keeping the two sides of the equation balanced, you can solve inequalities by keeping the two sides of the inequality unbalanced.

EXPLORE IMBALANCE

1. Open **Inequalities by Balancing.gsp.**

On page 1 the pans are balanced. Use this page to review the rules for what you can do without affecting the balance.

Q1 Drag a 1 from the storage area to each pan. Then drag a $-x$ to each pan. Do these steps disturb the balance of the pans? This is Rule 1. Write it down.

Q2 Drag an x and a -1 to each pan. Remove the x and $-x$ from the left pan. Remove the 1 and -1 from the right pan. When you remove two opposite objects from a pan, does it disturb the balance? This is Rule 2. Write it down.

Q3 On page 2 the pans are not balanced. Write down the inequality the pans represent.

As long as you follow the rules, you will not disturb the state of the pans.

Q4 Move objects on and off the pans (always following the two rules) until you get a single x all by itself on one pan and only numbers on the other pan. Write down this inequality.

Q5 Page 3 has a different arrangement, but the pans are still unbalanced. Use the two rules again to solve this inequality. Write down the original inequality, the steps you use, and the final inequality (when a single x is left on a pan by itself).

Rule 3 for inequalities is more limited than it is for equations. Be sure to use it to multiply or divide only by numbers that you know are positive.

Q6 On page 4 the pans are balanced again. This time the objects are arranged in two identical stacks on each pan. What happens if you remove one complete stack from each pan? What arithmetic operation does this correspond to? This is Rule 3. Write it down.

Q7 Page 5 has unbalanced pans. Write down the inequality. Use the rules to get a single x on one pan and only numbers on the other. Write down the solution.

EXPLORE MORE

You can adjust the value of x if you want. Press the *Show x* button and then move the slider.

Q8 Page 6 has empty pans. Create your own problem by dragging objects onto the pans. Make sure this is a problem that can be solved without fractions. Save your problem, and ask a classmate to try it.

Solving Compound Inequalities

SUBSTITUTION ON THE NUMBER LINE

Q1 The red segment appears above the number line if and only if the left (red) inequality is true for the current value of x. The blue segment appears below the number line if and only if the right (blue) inequality is true.

Q2 $x < 2$

Q3 $x > -3$

Q4 All real numbers are solutions to this compound inequality.

Q5 $x < 2$ and $x > -3$

This can be written in the concise form: $-3 < x < 2$.

Q6 a. $x > -7$ b. $x > 1$
c. $x < 0$ or $x > 6$ d. No solution.

SYMBOLIC SOLUTIONS

Q7 The filled circle at -4 indicates that this is at one end of the solution set, and this value satisfies the inequality. The open circle at 2 indicates that this too is at the end of the solution set, but it is not part of the solution.

5. Encourage students to experiment here. They cannot edit the numbers of the inequality directly, but they can change the smaller numbers above.

Q8 a. $x > -2$

b. All real numbers.

c. $x > 3$ and $x \leq 8$ ($3 < x \leq 8$)

THE INEQUALITY GAME

This is good practice for students who are learning these concepts for the first time. The game can continue indefinitely, and it does not require a great deal of guidance.

Each round of the inequality game requires a bit of time for calculations and thought, so it is not very fast-paced. Rather than competing against each other, it may be more effective to have students work in pairs and try to answer each challenge. Have them compare and discuss their own solutions before showing the correct answer.

WHOLE-CLASS PRESENTATION

Use **Compound Inequalities.gsp** with the Presenter Notes to present this activity to the whole class.

Solving Compound Inequalities

With compound inequalities, you may find that students' greatest problem is not solving the inequalities but correctly interpreting the and/or logical connectives. Page 1 of the Sketchpad document uses the trace feature to give a rough plot of the solutions to two inequalities. You and the class can then determine where at least one inequality is satisfied ("or") and where both are satisfied ("and").

1. Open **Compound Inequalities.gsp.** Explain to the class that the value controlled by point x is substituted into the expressions used on either side of the two inequalities.

Q1 Slowly drag point x along the number line. Tell the class to watch the red line segment above the point and the red inequality on the left. What condition determines whether the line segment will appear? (It appears if and only if x satisfies the inequality. There is a similar relationship for the blue segment that appears below.)

2. Select both of the line segments. Choose **Display | Trace Segments.** Drag point x to trace the solutions.

Q2 Derive the solutions to both inequalities symbolically in order to demonstrate that the traces match the solutions. Ask the class for the range of numbers on which at least one of the inequalities is satisfied (all real numbers), and ask them where both are satisfied ($-3 < x < 2$). Explain to them that these are the solutions to the compound inequalities below:

$$2x < 6 - x \quad \text{or} \quad 7x + 10 > x - 8$$

$$2x < 6 - x \quad \text{and} \quad 7x + 10 > x - 8$$

Remember to use the action buttons to change the directions of the inequality signs when necessary.

Q3 Edit the calculations and use the same procedure to derive the solutions to the compound inequalities below. Press the *Erase Traces* button, and drag point x.

$$13 - x > -5x - 15 \quad \text{or} \quad 4x + 10 > 17 - 3x \qquad (x > -7)$$

$$13 - x > -5x - 15 \quad \text{and} \quad 4x + 10 > 17 - 3x \qquad (x > 1)$$

$$30 - 13x > 30 - 7x \quad \text{or} \quad 8x + 18 < 11x \qquad (x < 0 \text{ or } x > 6)$$

$$30 - 13x > 30 - 7x \quad \text{and} \quad 8x + 18 < 11x \qquad (\text{No solution.})$$

Go to the Compound Inequality page to play an inequality game. This is a bit more sophisticated. It handles six relationships ($<, >, \leq, \geq, =, \neq$) and plots the solution to a combination of two linear inequalities (or equations). Press *Play* to generate two inequalities randomly. Give students time to sketch the solution. Press *Show* to reveal the solution on the number line.

Exploring Algebra 1 with The Geometer's Sketchpad
© 2012 Key Curriculum Press

Solving Compound Inequalities

A *simple inequality* declares one condition that must be true. For example, to satisfy the inequality $2x < 3 + x$, $2x$ must be less than $3 + x$. A *compound inequality* declares two or more conditions, and some combination of them must be true. For this activity, we will stick to two inequalities.

SUBSTITUTION ON THE NUMBER LINE

1. Open **Compound Inequalities.gsp.**

Point x is attached to the number line, and its coordinate is used to evaluate the two inequalities above the number line:

$$2x < 6 - x \qquad 7x + 10 > x - 8$$

Q1 Drag point x along the number line. A line segment sometimes appears above the points, and another one sometimes appears below. For each line segment, what determines whether it will appear?

2. Select both line segments. Choose **Display | Trace Segments.**

3. Slowly drag point x to trace the solutions of both inequalities.

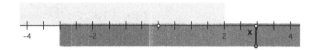

Q2 What is the solution set of the left (red) inequality?

Q3 What is the solution set of the right (blue) inequality?

Q4 Consider this compound inequality:

$$2x < 6 - x \quad \text{or} \quad 7x + 10 > x - 8$$

> When the word "or" is used, the solution is the *union* of the two sets.

The word "or" indicates that the solution includes all values of x for which one or both inequalities are true. What is the solution set of this compound inequality?

Q5 This is a different compound inequality:

$$2x < 6 - x \quad \text{and} \quad 7x + 10 > x - 8$$

> When the word "and" is used, the solution is the *intersection* of the two sets.

The word "and" indicates that the solution includes only the values of x for which both inequalities are true. What is the solution set of this compound inequality?

Q6 Model the compound inequalities on the next page, and report their solution sets. To change the direction of an inequality sign, click the button above the sign. To edit one of the four expressions, double-click it; to enter x into the calculation, click the measurement x in the sketch. After you set the inequality signs and expressions correctly, press the *Erase Traces* button, and drag x to see the new trace.

a. $13 - x > -5x - 15$ or $4x + 10 > 17 - 3x$

    b. $13 - x > -5x - 15$ and $4x + 10 > 17 - 3x$

    c. $30 - 13x > 30 - 7x$ or $8x + 18 < 11x$

    d. $30 - 13x > 30 - 7x$ and $8x + 18 < 11x$

SYMBOLIC SOLUTIONS

The method you have been using amounts to guess-and-check. It's more reliable and often more efficient to solve both inequalities symbolically by undoing operations, and then to compare the solution sets.

4. Page 2 shows a number line plot of the compound inequalities $x \geq -4$ and $x < 2$. Notice that the inequality is already solved.

Q7 Why is the circle at -4 filled while the one at 2 is open?

5. Take a few minutes to experiment with the objects on this page. Press the buttons and edit the small numbers above the inequalities.

Q8 To solve the compound inequalities below, solve the parts separately and combine the solutions using the sketch. In each case, report the solution set and sketch a number line plot.

    a. $6x < 9x - 30$ or $4x + 12 > -x + 2$

    b. $5x \leq 8x + 24$ or $7x - 10 < 2x + 15$

    c. $5x - 9 > 12 - 2x$ and $3x + 13 \leq 45 - x$

THE INEQUALITY GAME

The One Inequality page shows an inequality (or sometimes an equation). Press the *Play* button to change it. Then work the graph out by yourself and press the *Show* button to check your answer.

Go to the Compound Inequality page. Press the *and/or* button to choose which type of compound inequality to use.

The One Graph and Compound Graph pages provide the same games in reverse: You see the graph and must derive the expression. On the Compound Graph page, if your answer does not match the one on the screen, check again. You may be right. There is often more than one expression that will produce the same graph.

5

Coordinates, Slope, and Distance

Coordinates: The Fly on the Ceiling

Students learn about rectangular coordinates by investigating Descartes' original inspiration for the system, a fly walking on a bedroom ceiling. They measure coordinates and plot points.

The Origin: Center of the World

Students investigate a coordinate system based on an origin, use both positive and negative coordinates, identify the four quadrants, and use coordinates to draw figures.

Points Lining Up in the Plane

Students are informally, experientially introduced to the relationship between descriptions of coordinate patterns and graphs in the Cartesian plane. Too often students don't really understand the connection between an equation and its graph. This activity fosters the understanding that graphs depict the set of points whose coordinates satisfy an equation.

The Slope of a Line

Students explore slope as a measure of steepness. The purpose of the activity is to provide a general feel for slope before students are introduced to how slope is calculated. The focus is on qualitative relationships such as how all lines with a positive slope are different from all lines with a negative slope.

The Slope Game

Students acquire an intuitive feel for slope as they construct and play a game in which one player rearranges lines on the screen and the other player tries to match each line with its slope measurement. The game can be modified to play alone as well.

More Slope Games

Students acquire an intuitive feel for slope by competing to associate lines with their corresponding slopes. Practicing in a fun environment, students build an inventory of mental reference images: It does not take long to visualize lines with slopes of 1, 0, or −1.

How Slope Is Measured

Students appreciate the problem of how to measure steepness in the real-world context of building a staircase, and connect intuitions about slope to specific calculations based on the coordinates of two points (the slope formula).

Slopes of Parallel and Perpendicular Lines

Students explore the relationship between slopes of parallel or perpendicular lines.

The Pythagorean Theorem

Students verify the Pythagorean theorem by examining a right triangle plotted on a square grid. They then apply the theorem on the coordinate grid and develop the distance formula.

Exploring Algebra 1 with The Geometer's Sketchpad
© 2012 Key Curriculum Press

Coordinates: The Fly on the Ceiling

DESCARTES' FLY

1. Make sure students understand that the numbers in the table correspond to the metric measurements from the two walls to the point.

2. Here we introduce x and y as general names for the coordinates.

3. A class discussion based on the coordinate pairs of the four corners of the rectangle is a fine way to ensure that all students have understood the concept before moving on.

Q1 The room is 6 meters long and 4 meters wide. These are the ranges of the x- and y-coordinates.

Q2 The coordinates of the center of the rectangle are $(3, 2)$.

Q3 The coordinates of the point are $(4, 1)$.

PLOTTING POINTS

The amount of detail in students' geometric descriptions will depend on their geometry background.

Q4 The figure is a square with side length 2 m. Its perimeter is 8 m and its area is 4 m^2.

Q5 The figure is an isosceles trapezoid. The bases are 2 m and 4 m. The two other sides are each $\sqrt{5}$ m. The perimeter is $(6 + 2\sqrt{5})$ m and the area is 6 m^2.

Q6 The missing coordinates are $(1, 3)$ and $(3, 1)$.

Q7 There are many pairs of points that can be used to make the rhombus. The coordinates must be in this general form: $(3 - a, 2)$ and $(3 + a, 2)$, where $0 \leq a \leq 3$. Here are the possible solutions using integers only:

$(0, 2)$ and $(6, 2)$

$(1, 2)$ and $(5, 2)$

$(2, 2)$ and $(4, 2)$

 ACTIVITY NOTES

EXPLORE MORE

Q8 The constellation is the Big Dipper, one of the most distinctive constellations in the northern sky.

Coordinates: The Fly on the Ceiling

René Descartes, it is said, first got his idea for the rectangular coordinate system while he was pondering a fly that was walking on a ceiling. He knew that he could fix the fly's position in the plane with two numbers, the distances to two adjacent walls. Give students a clear picture of this situation, and perhaps some historical context.

1. Open **Coordinates Present.gsp.**

2. Drag the red point around the interior of the rectangle. Explain that the point represents the fly and the rectangle is the ceiling.

Q1 Ask what the two distances are measuring. They are the distances from the left edge and from the bottom edge.

3. Draw students' attention to the table at the right. These are the same distances written as coordinates. It is important that the students know which number is which.

Q2 Is it possible for one point to have more than one pair of coordinates? Is it possible for two different points to have the same pair of coordinates?

4. Drag the point to a corner of the ceiling. Double-click the table. This will save the coordinates of that point as a row in the table. Repeat at the other corners.

Q3 Ask the students to tell you the length and width of the ceiling (6 m by 4 m).

Q4 Ask what the coordinates would be at the very center of the room (3, 2).

5. Show how to plot points by coordinates. Choose **Graph | Plot Points.** Enter 3 and 2. Click Plot, then Done.

6. Go to page 7 and press *Show Constellation.* You will see an image of the constellation Leo. Drag the point to each star in turn, and double-click the table. After the last one, hide the constellation.

Explain that you have now recorded what you saw, and you can use the coordinates to reproduce the same image. In fact, anyone anywhere could do the same, even if they had never seen Leo.

7. Select the table. Choose **Graph | Plot Table Data.**

A good follow-up would be to have students draw patterns on graph paper, record the coordinates, and have someone else reproduce the image from the coordinates.

Q5 Ask students to name practical applications of coordinates that they have already used. Tell them that many digital computer images (GIFs, JPEGs) are just more sophisticated versions of the recording technique you used with the star constellation.

Coordinates: The Fly on the Ceiling

Descartes is perhaps better known as a philosopher than as a mathematician. His most famous quote is *Cogito ergo sum*—"I think, therefore I am."

As the story goes, philosopher and mathematician René Descartes was gazing upward, deep in thought, when he saw a fly walking on the ceiling. It occurred to Descartes that he could describe the fly's position on the ceiling by two numbers: its distance from each of two walls. Thus was born the *coordinate plane,* also called the *Cartesian coordinate system* after Descartes.

In this activity you'll investigate the original idea of Descartes.

DESCARTES' FLY

Open **Coordinates.gsp.** You will see a model of the ceiling in Descartes' bedroom. Imagine that the red point is the fly in the story.

You can move the selected point in very small steps using the arrow keys on the keyboard.

1. Move the point and notice how the measurements change on the sketch and in the table to the right. These measurements are called *coordinates.*

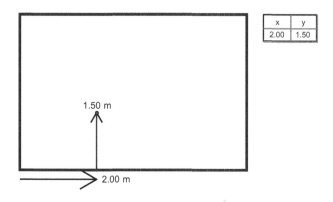

2. Move the point to the upper right corner. Double-click the table to enter the measurements of the *x*- and *y*-coordinates.

3. Move the point to the other three corners and enter their coordinates into the table. Compare your four pairs of measurements with the measurements of other students.

Q1 How long and wide is the room? Explain how you can figure this out from the coordinates of the four corners.

4. Place the point as close as possible to the center of the room by eye.

Q2 What are the coordinates for the center point? Describe how you can figure this out using the coordinates of the four corners.

Q3 What are the coordinates for a point $\frac{1}{4}$ from the bottom and $\frac{2}{3}$ from the left?

Exploring Algebra 1 with The Geometer's Sketchpad
© 2012 Key Curriculum Press

PLOTTING POINTS

In this section you will plot points using their coordinates.

5. On page 2, plot these four points: $(1, 1)$, $(1, 3)$, $(3, 1)$, and $(3, 3)$. To plot them, choose **Graph | Plot Points.** For each point, enter its coordinates and click Plot. Click Done after you finish the last point.

Q4 Describe in detail the figure outlined by the points.

Q5 On page 3, select the table and choose **Graph | Plot Table Data.** Click **OK.** Describe the figure outlined by the points.

You can find the coordinates of a point by selecting it and choosing **Measure | Coordinates.**

Q6 Page 4 shows two points that are opposite vertices of a square. What are the coordinates for the two missing vertices? Plot them.

Q7 The two points on page 5 are opposite vertices of a rhombus. Write down three pairs of possible coordinates for the two missing points. Plot the missing points.

Carpenters and other craftsmen use coordinates to mark panels for ceilings and walls so they can find water pipes and electrical wires hidden behind the panels. Before they cover a wall, they take notes on the coordinates for such hidden objects. Afterwards they can cut the necessary holes to install plumbing fixtures and electrical outlets and switches.

EXPLORE MORE

6. On page 6 there is a table with a set of coordinates. You will use them to decorate the ceiling in Descartes' room. Each ordered pair represents the position of a star in a well-known constellation.

Q8 Plot the table data. What is the name of the constellation?

Q9 Find a picture of your favorite constellation. Frame the picture in a rectangle, and measure the coordinates of each star with a metric ruler. Use page 7 of the document to plot the data one by one and make your own ceiling decoration.

You can start with a digital picture from the web. Use a search engine to find images.

Q10 You can even do this directly on the screen with Sketchpad. Copy any digital picture and paste it into page 7, or use the picture already in the sketch. Record in the table the positions of the edges of the objects in the picture. When you are done, hide the picture and plot the table data.

The Origin: Center of the World

CENTER OF THE WORLD

Q1 When the point is at a position lower than the origin, the y-coordinate is negative. When the point is to the left of the origin, the x-coordinate is negative.

Q2 The rectangle is 8 units long and 5 units wide. Find the length by subtracting the x-coordinate of a point on the left from the x-coordinate of a point on the right. Similarly, find the width by calculating the difference between the y-coordinates of points on the top and bottom.

Q3 The size of the rectangle is independent of the position of the origin. Thus the calculation should be the same as in Q2 even if the measured coordinates are different.

BREAKING THE WALLS

Q4 The dimensions of the rectangle should still be the same. Changing the position of the origin point makes no difference at all.

Q5 In Quadrant I, both coordinates are positive.

In Quadrant II, x is negative and y is positive.

In Quadrant III, both coordinates are negative.

In Quadrant IV, x is positive and y is negative.

Q6 If the origin is far to the southwest, all of the coordinates at the project site will be positive. This simplifies the calculations.

INVESTIGATE MORE

Q7 For this task, it is best to evaluate student progress by watching them as they work. Have them exchange coordinates to see if they can duplicate each other's work.

This is the final test. If students can draw these figures using coordinates, they probably have a solid understanding of the basics regarding the coordinate system.

Exploring Algebra 1 with The Geometer's Sketchpad
© 2012 Key Curriculum Press

 ACTIVITY NOTES

IF TIME PERMITS

Students can finish this activity by working in pairs. First they should decide on a figure. It could be a hexagon, an arrow, or something else of their choice. They then discuss and decide which coordinates they should use to outline the figure. Finally, they can test their guess using the premade Target sketch.

The Origin: Center of the World

This activity is a follow-up to Coordinates: The Fly on the Ceiling. Rather than measure the distances from two walls, you will use a single reference point, the origin.

1. Open **The Origin Present.gsp.** The biggest difference from the earlier activity is that the coordinates are now measured from an origin point in the middle of the rectangle, rather than using two walls for reference.

Q1 Drag point *A*, but keep it above and to the right of the origin. Ask the class what they expect to happen when you drag it to the left of the origin. The *x*-coordinate will become negative. Show this slowly. Show that the *y*-coordinate becomes negative when you drag it below the origin.

Q2 Tell the class that you want to measure the dimensions of the rectangle, and ask them for guidance. You can get the length by dragging point *A* to the right border, then the left, and comparing the *x*-coordinates. Check the *y*-coordinates at the top and bottom to get the width. To record coordinates of a point, drag point *A* into place and double-click the table.

2. Move origin point *O* to another location and repeat the measurements. Move it out of the rectangle and do the measurements once more. Check that students understand that changing the location of the origin will not change the measurements.

3. Press the *Hide Rectangle* and *Show Coordinate System* buttons. Review the names of the two axes and the four quadrants.

Q3 In what quadrant must point *A* be if both coordinates are negative? [III] Discuss the connections between the signs of the coordinates and the four quadrants.

4. Open the Target page of the document. The two coordinates control the position of the target (the cross). Press the *Go to Target* button. The arrow will travel to the target, leaving a trail as it moves. Demonstrate it by drawing a square with this sequence of points: (1, 1), (5, 1), (5, 5), (1, 5), (1, 1).

Q4 Get students to help you with coordinate plots of other figures. Here are some suggestions:

   a. right triangle

   b. isosceles triangle

   c. rhombus

   d. trapezoid

   e. star

The Origin: Center of the World

The word *origin* comes from Latin and means "place of birth."

In the activity Coordinates: The Fly on the Ceiling we see how Descartes measured the coordinates of a fly by noting its distance from each of two walls. It's possible to use instead a single fixed point (such as a light hanging in the middle of the room) as a reference point. We call this reference point the *origin*, and we use it as our chosen "center of the world."

CENTER OF THE WORLD

You can move the selected point in very small steps using the arrow keys on the keyboard.

Open **The Origin.gsp.** The enlarged point near the center of the sketch is the origin. You can describe the position of any other point by using the horizontal and vertical measurements from the origin.

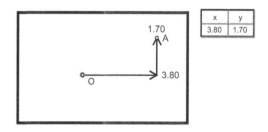

x	y
3.80	1.70

1. Move point *A* and notice how the measurements change on the sketch and in the table to the right.

Q1 What happens to the *y*-coordinate when the position of the point is lower than the origin? And what happens to the *x*-coordinate when the point is to the left of the origin?

Add values to a table by double-clicking the table.

2. Move the point until it touches the right side of the rectangle. Add the *x*- and *y*-coordinate measurements to the table.

3. Do the same with the other three sides. Be careful to remember which row of the table corresponds to each side.

Q2 Use your measurements to find the length and width of the rectangle. Explain how you got your results.

Q3 Move the origin point to a new position, and use the same procedure to measure the dimensions of the rectangle. Did the size of the rectangle change? Explain your result.

BREAKING THE WALLS

4. Move the origin outside of the rectangle and use coordinates to measure the dimensions again.

Q4 Did you find any difference? When using coordinates to measure distances, what difference does the location of the origin make?

5. Press the *Show Coordinate System* button and the *Hide Rectangle* button.

The *x*- and *y*-axes divide the plane into four regions called quadrants, numbered I, II, III, and IV, as shown here.

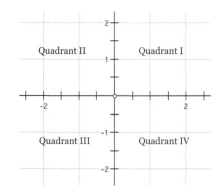

Always refer to quadrants with Roman numerals.

The axes are boundary lines and do not lie in any of the quadrants.

Q5 If a given point is in Quadrant I, what is the sign of its *x*-coordinate? What is the sign of its *y*-coordinate? For each of the quadrants, state a general rule about the signs of the coordinates of a point in that quadrant.

Q6 The origin for a surveying or mapping project is usually a point far away in a southwesterly direction. What do you suppose is the reason for choosing such a point for the origin?

INVESTIGATE MORE

You can change a parameter by double-clicking it, or you can select it and press the + and – keys.

6. Open the Target page of the same document. You can move the target (shown as a black cross) by changing the *x* and *y* parameters. Then move the arrow and leave a trace by pressing the *Go to Target* button. The arrow will go to the target, tracing its path as it does so. Practice changing the parameters and moving the arrow until you get a feel for the mechanics in the sketch.

To erase the traces, choose **Display | Erase Traces.**

Q7 Erase all traces before working on each of the following exercises. In each case, record the coordinates you use.

    a. Draw a square.

    b. Draw a triangle.

    c. Draw a rhombus.

    d. Make your own figure. It could be a star, a maze, or something different.

Exploring Algebra 1 with The Geometer's Sketchpad
© 2012 Key Curriculum Press

Points Lining Up in the Plane

The purpose of this activity is to give students an informal and experiential introduction to the relationship between descriptions of coordinate patterns and graphs in the Cartesian plane. Too often, students don't really get the connection between an equation and its graph. It's important for them to understand that graphs depict the set of points whose coordinates satisfy an equation. This activity helps foster that understanding.

To deepen the experience, conduct a class or group discussion that encourages students to ponder this relationship. Ask, "Why do the points 'line up' in such regular ways? If you could plot not just five, but every point that satisfies the description, what would that look like?"

SKETCH AND INVESTIGATE

Q1 In each case, the answer shown depicts all possible answers with integer coordinates on the grid provided. The question asks for five answers, so any five of the points shown is a correct response (not to mention the infinite number of correct responses outside the grid!).

a.

b.

c.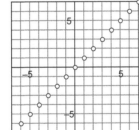

d.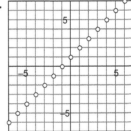

e.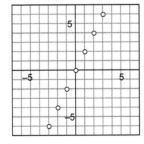

f.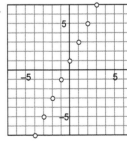

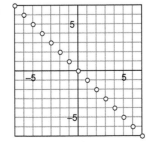

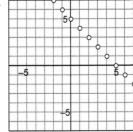

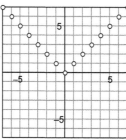

g.   h.

Q2 a. The *y*-coordinate equals the *x*-coordinate.

b. The *y*-coordinate is one less than the *x*-coordinate.

c. The *y*-coordinate is twice the *x*-coordinate. (Or, the *x*-coordinate is one-half the *y*-coordinate.)

d. The *y*-coordinate is two less than twice the *x*-coordinate.

e. The *y*-coordinate is one-third the *x*-coordinate. (Or, the *x*-coordinate is three times the *y*-coordinate.)

f. The *y*-coordinate is always −1 (regardless of the value of the *x*-coordinate).

g. The *y*-coordinate is the opposite of the absolute value of the *x*-coordinate. (An acceptable alternate answer for students not familiar with the term *absolute value* might be "The *y*-coordinate is the 'negative value' of the *x*-coordinate, regardless of whether the *x*-coordinate is positive or negative.")

h. The product of the *y*-coordinate and the *x*-coordinate is 6.

EXPLORE MORE

Q3 Equations from Q1:

a. $y = x$ b. $y = x + 1$ c. $y = 2x$

d. $y = 2x + 1$ e. $y = -x$ f. $x + y = 5$

g. $y = |x|$ h. $y = x^2$

Equations from Q2:

a. $y = x$ b. $y = x - 1$ c. $y = 2x$

d. $y = 2x - 2$ e. $y = (1/3)x$ or $x = 3y$

f. $y = -1$ g. $y = -|x|$ h. $xy = 6$

Q4 Answers will vary.

Here's how to set up the *Movement* button (more detailed instructions are on page 2 of **Points Line Up.gsp**): Plot the eight destination points using the **Plot Points** command. Select all 16 points in the sketch in the following order: point *P*, point *P*'s destination, point *Q*, point *Q*'s

destination, point *R*, point *R*'s destination, . . . , point *W*, point *W*'s destination. Now choose **Edit | Action Buttons | Movement.** Change the speed and label (on the Label panel), and then click OK. Now hide the eight destination points (using **Display | Hide**).

Q5 Answers will vary, but should line up with answers to Q1 and satisfy the rule.

WHOLE-CLASS PRESENTATION

Students connect verbal and graphical representations of points by using a verbal rule about coordinates to position points and by observing a pattern of points to formulate a verbal rule about their coordinates.

Use the sketch **Points Line Up Present.gsp** in conjunction with the Presenter Notes to present this activity to the whole class.

It's best to have a different student volunteer operate the computer for each rule.

1. Open **Points Line Up Present.gsp.**

2. Drag point P so that students can see how the coordinates change. Explain that students will take turns dragging the points to make the coordinates satisfy certain rules.

Encourage students to help each other in figuring out how to move the points and formulate rules.

3. Have the first student volunteer press button a to show the first rule, read the rule out loud, and then drag point P around until it satisfies the rule.

4. Have the student drag each of the remaining points around until all the points satisfy the rule.

Q1 Ask the class, "How would you describe the pattern these points make?"

Q2 Ask students to record on their paper both the rule and a diagram showing how the points are arranged.

5. Have a second student volunteer come to the computer, press button b, and drag the points for the second rule. Have students record on their paper each rule and a diagram of the resulting pattern. Continue for as many of the remaining rules as seems appropriate.

6. Go to page 2 of the sketch, and explain that on this page the points will arrange themselves and that the job of the class is to make up a rule that fits.

7. After students have written rules for all the arrangements on page 2, press the a button again to return the points to their initial arrangement.

8. Ask, "What rule did you write down for this arrangement?" After students have responded, ask "Does anyone know how to write this rule as an equation?" Make sure students understand why the answer is $y = x$.

9. Press each of the remaining buttons in turn, and have students give the equation for each pattern.

10. Tell students to go back to their answers for page 1 and write an equation for each of those arrangements.

Finish with a class discussion encouraging students to describe their insights. The discussion might consider questions such as these:

"How many points are there that would satisfy one of these rules?"

"If you could plot all the points that satisfy a rule, what would the result look like?"

"Why is it that the points line up so neatly?"

Points Lining Up in the Plane

If you've seen marching bands perform at football games, you've probably seen band members wandering in seemingly random directions suddenly spell a word or form a cool picture. Can you describe these patterns mathematically? In this activity you'll start to answer this question by exploring simple patterns of dots in the *xy* plane.

SKETCH AND INVESTIGATE

1. In a new sketch, choose the **Point** tool from the Toolbox.

Holding down the Shift key keeps all five points selected.

2. While holding down the Shift key, construct five points.

3. With all points selected, choose **Display | Label Points.** Set the label of the first point to *P* and click OK.

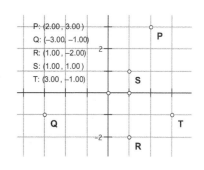

To measure the coordinates, choose **Measure | Coordinates.**

4. Measure the coordinates of the five selected points.

A coordinate system appears, and the coordinates of the five points are displayed.

To hide objects, select them and choose **Display | Hide.**

5. Hide the point at the origin $(0, 0)$, and the unit point $(1, 0)$.

6. To make dragged points land only on locations with integer coordinates, choose **Graph | Snap Points.**

Q1 For each part a–h below, drag the five points to five different locations that satisfy the rule. Then copy your solutions onto the grids on the next page. Remember to fill in the coordinates of each point.

   a. The *y*-coordinate equals the *x*-coordinate.

   b. The *y*-coordinate is one greater than the *x*-coordinate.

   c. The *y*-coordinate is twice the *x*-coordinate.

   d. The *y*-coordinate is one greater than twice the *x*-coordinate.

   e. The *y*-coordinate is the opposite of the *x*-coordinate.

   f. The sum of the *x*- and *y*-coordinates is 5.

The absolute value of a number is its "positive value." The absolute value of both 5 and −5 is 5.

   g. The *y*-coordinate is the absolute value of the *x*-coordinate.

   h. The *y*-coordinate is the square of the *x*-coordinate.

a. The *y*-coordinate equals the *x*-coordinate.
   P: (   )
   Q: (   )
   R: (   )
   S: (   )
   T: (   )

b. The *y*-coordinate is one greater than the *x*-coordinate.
   P: (   )
   Q: (   )
   R: (   )
   S: (   )
   T: (   )

c. The *y*-coordinate is twice the *x*-coordinate.
   P: (   )
   Q: (   )
   R: (   )
   S: (   )
   T: (   )

d. The *y*-coordinate is one greater than twice the *x*-coordinate.
   P: (   )
   Q: (   )
   R: (   )
   S: (   )
   T: (   )

e. The *y*-coordinate is the opposite of the *x*-coordinate.
   P: (   )
   Q: (   )
   R: (   )
   S: (   )
   T: (   )

f. The sum of the *x*- and *y*-coordinates is 5.
   P: (   )
   Q: (   )
   R: (   )
   S: (   )
   T: (   )

g. The *y*-coordinate is the absolute value of the *x*-coordinate.
   P: (   )
   Q: (   )
   R: (   )
   S: (   )
   T: (   )

h. The *y*-coordinate is the square of the *x*-coordinate.
   P: (   )
   Q: (   )
   R: (   )
   S: (   )
   T: (   )

Exploring Algebra 1 with The Geometer's Sketchpad
© 2012 Key Curriculum Press

BACKWARD THINKING

In Q1, you moved points around to make them fit a certain rule. Here you'll reverse the process and be the detective. Your clues will be the given point positions, and your task will be to figure out the rule.

7. Open **Points Line Up.gsp.** You'll see a coordinate system with eight points (*P–W*), their coordinate measurements, and eight action buttons (*a–h*).

Q2 Press each action button in the sketch. Like the members of a marching band, the points will move about until they form a pattern. Study the coordinates of the points in each pattern, and then write a rule (like the ones in Q1) for each of the action buttons *a–h*.

EXPLORE MORE

Q3 Each rule in this activity can be written as an equation. For example, the rule for part b of Q1 ("The *y*-coordinate is one greater than the *x*-coordinate.") can be written as $y = x + 1$. Write an equation for each description in Q1 and Q2.

Q4 Add your own action buttons to those in **Points Line Up.gsp,** and ask your classmates to come up with descriptions or equations for your patterns. (Instructions on how to do this are on page 2 of the sketch.)

Q5 Go back to your first sketch. Turn off **Graph | Snap Points.** Drag each of your five points to a new location that satisfies the rule without using integer coordinates. Add your new points to grids a–h, and write their coordinates down on your paper.

The Slope of a Line

The purpose of this activity is to get a general feel for slope before learning exactly how slope is calculated. The focus should be on qualitative relationships such as how all lines with a positive slope are different from all lines with a negative slope.

A good way to check if students have really internalized the concepts in this activity is to have them model lines using their forearms. Ask the class which arm would be more convenient for modeling lines with a positive slope (left) and which for those with a negative slope (right). Start by having students model slopes of 1 and −1, and remind them that these should be thought of as "points" of reference. Then call out slope values ("5 . . . −1 . . . 0.2 . . . −1/2 . . . 0 . . . −100"), and have students quickly approximate them with their arms.

An interesting discussion topic during or after this activity is whether there is a biggest or a smallest possible slope for a line.

SKETCH AND INVESTIGATE

3. The labels of the points should appear when students measure the slope. If not, students can click the **Text** tool on the points to show the labels. (You can choose **Edit | Preferences** and use the Text panel to control whether to show labels automatically as objects are measured. The normal setting is to show labels.)

Q1 a. A line with a slope of 1 will go up to the right and make angles of 45° with both axes.

b. A line with a slope of −1 will go up to the left and will also make angles of 45° with both axes.

c. A line with a slope of 0 is perfectly flat—horizontal, in other words.

d. A line with an undefined slope is perfectly vertical.

e. Lines with positive slopes always go up to the right and down to the left, regardless of how steep they are.

f. Lines with negative slopes always go up to the left and down to the right, regardless of how steep they are.

Q2 Lines with a slope greater than 1 are steeper than lines with a slope of 1. Lines with a slope between 0 and 1 are less steep than lines with a slope of 1. (It's helpful to think of lines with a slope of 1 as the "middle case" between steeper and less steep.)

Q3 They are just as steep, but in the opposite direction (one goes up to the right, the other up to the left). A fancy way of expressing this is that the two lines are reflections of each other across the vertical line through their point of intersection.

Q4 The line is translated—in other words, shifted in some direction—but not rotated, so its slope doesn't change, and neither does the slope measurement. One way of saying this is that the line is dragged "parallel to itself."

Q5 There are infinitely many solutions for each blank in the table. (Why?) Those listed here are the ones that fit in a normal-sized sketch window. Where fewer than three points fit, the three closest points—one in either direction—are listed.

A	B	Slope $\overleftrightarrow{AB}$
(0, 0)	(1, 2), (2, 4), (−1, −2), or (−2, −4)	2.000
(2, 3)	(3, 0), (4, −3), (1, 6), or (5, −6)	−3.000
(−1, 4)	(−2, 4), (0, 4), (1, 4), (2, 4), or (3, 4)	0.000
(2, −5)	(6, 0), (10, 5), or (−2, −10)	1.250
(3, 1)	(3, −2), (3, −1), (3, 0), or (3, 2)	Undefined
(−1, −3)	(−4, 5), (−7, 13), or (2, −11)	−2.667
(−3, 2)	(5, 3), (13, 4), or (−11, 1)	0.125
(4, 2)	(2, −5), (0, −12), or (6, 9)	3.500

EXPLORE MORE

Q6 Answers will vary. There are many possible relationships that students may notice. Some of these relationships are specific to particular slopes and positions of *A*. For instance, when *A* is at (0, 0) and the slope is 2, *B*'s *y*-coordinate is always twice its *x*-coordinate. At this early stage of exploring slope, it's more important to encourage careful observations than to develop the formal definition.

The effort to predict a new position of *B* from their observations helps students focus those observations.

Q7 The slope doesn't change when *A* and *B* exchange places, because they still determine the same line.

Q8 There are infinitely many non-integer coordinates for *B*. The points may follow similar patterns to the ones students observed in Q6, depending on what patterns they described.

WHOLE-CLASS PRESENTATION

Introduce or revisit slope as a measure of steepness, and ask students to think about how they might assign a numeric value to describe it. Use the sketch **Slope of a Line Present.gsp,** and drag *A* and *B* to show how slope changes. Pose questions Q1–Q5 from the activity and elicit student participation. Using **Graph | Snap Points** in step 5 restricts the coordinates of *A* and *B* to integer values, making it easier for students to see how slope is mathematically related to these numbers.

The Slope of a Line

The *steepness* of things—ski runs, wheelchair ramps, or lines in the *xy* plane—can be described in lots of ways. For instance, skiers know that "black diamond" runs are steep and challenging, whereas "green circle" runs are less steep and easier. Mathematicians prefer to use numbers to describe steepness so that they can compare the steepness of objects and solve problems. In this activity you'll explore *slope*, a number that describes a line's steepness.

SKETCH AND INVESTIGATE

1. In a new sketch, press and hold the **Straightedge** tool in the Toolbox. Choose the **Line** tool from the menu that pops out.

2. Draw a line. Measure its slope by choosing **Measure | Slope.**

3. Measure the coordinates of the two points *A* and *B* that define the line.

Deselect all objects by clicking in blank space with the **Arrow** tool. Then select the two points and choose **Measure | Coordinates.**

4. Drag *A* and *B* to different locations and observe the changes in the slope measurement of the line.

Q1 Describe lines that have these slopes:

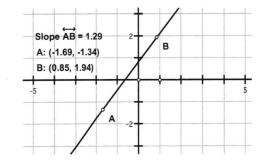

a. a slope of 1

b. a slope of −1

c. a slope of 0

d. an undefined slope

e. any positive slope

f. any negative slope

Q2 How do lines with a slope greater than 1 compare to lines with a slope of 1? How do lines with a slope between 0 and 1 compare to lines with a slope of 1?

Q3 Draw a new line. Drag it so that its slope is the opposite of the slope of your first line. How do lines having opposite slopes compare?

Be sure to drag the line itself and not either of its control points.

Q4 How does a line move when you drag it with the **Arrow** tool? What happens to the slope measurement when you drag a line this way?

5. Choose **Graph | Snap Points.** From now on, when you drag points, they will move only to locations with integer coordinates.

6. Display the slope to the nearest thousandth by selecting it and choosing **Edit | Properties.** Go to the Value tab and change the precision to **thousandths.**

Select the three
measurements in order
and choose **Number |
Tabulate.**

7. Create a table for the coordinates of *A*, the coordinates of *B*, and the slope.

Q5 The table below shows different locations of *A* and values for the slope of $\overleftrightarrow{AB}$. Move point *A* to the indicated location; then find the coordinates of two locations for *B* that make the slope of $\overleftrightarrow{AB}$ equal the value in the last column. Each time you find a good location for *B*, double-click the table to "lock in" your entry.

A	B	Slope $\overleftrightarrow{AB}$
(0.00, 0.00)	(2.00, 4.00)	2.000

You may need to make
your sketch window
larger to have enough
space to find two
answers to some of
these problems.

If you have access
to a printer, you can
print out your table by
printing the sketch.

A	*B*	*Slope $\overleftrightarrow{AB}$*
(0.00, 0.00)		2.000
(0.00, 0.00)		2.000
(2.00, 3.00)		−3.000
(2.00, 3.00)		−3.000
(−1.00, 4.00)		0.000
(−1.00, 4.00)		0.000
(2.00, −5.00)		1.250
(2.00, −5.00)		1.250
(3.00, 1.00)		Undefined
(3.00, 1.00)		Undefined
(−1.00, −3.00)		−2.667
(−1.00, −3.00)		−2.667
(−3.00, 2.00)		0.125
(−3.00, 2.00)		0.125
(4.00, 2.00)		3.500
(4.00, 2.00)		3.500

EXPLORE MORE

Q6 What is the relationship between the slope and the coordinates of *A* and *B*? Write down any patterns you notice, and use them to predict a third possible location for *B* for each slope and point *A* in the table. Use Sketchpad to check your prediction.

Q7 What happens if *A* and *B* trade places? Pick a row in the table and switch the locations of *A* and *B*. What happens to the slope? Why?

Q8 Pick any row in the table above. Turn off **Graph | Snap Points** and find some non-integer coordinates for *B*. Does this change your answer to Q6? How?

Exploring Algebra 1 with The Geometer's Sketchpad
© 2012 Key Curriculum Press

The Slope Game

This simple, unassuming game has been a favorite in classrooms and workshops for years. Students really do enjoy trying to trick each other with lines that are very close to each other in slope, or the opposite of each other, and this represents a good learning opportunity.

PLAYING THE SLOPE GAME

2. This step says to draw "five different random lines." The lines should not be attached to each other. In other words, students should click or release only in blank space when constructing the lines so that all the control points are independent points.

AFTER PLAYING

The More Slope Games activity contains four different ready-made slope games that can be played in a variety of ways. They can be played immediately after this game, but we recommend spreading them out over several days or weeks, letting students play each one for 5–15 minutes as time permits.

The Slope Game

Imagine a game that combines the best elements of laser tag, Doom, and chess. This is not that game, but it is still a fun math game that's good for solidifying your sense of slope. The game works best with a partner (this is how it's described), but you can also play it alone if you hide the labels and cover the slope measurements before dragging the lines.

PLAYING THE SLOPE GAME

1. In a new sketch, press and hold the **Straightedge** tool in the Toolbox. Choose the **Line** tool from the menu that pops out.

Click and release only in blank space so that none of the lines are attached to each other.

2. Draw five different random lines in your sketch.

3. Measure the slopes of the five lines by selecting all of them at once and choosing **Measure | Slopes.**

An easy way to select all the points is to choose the **Point** tool and then choose **Edit | Select All Points.**

4. Create a *Hide/Show* action button by selecting all the points and choosing **Edit | Action Buttons | Hide/Show.**

5. Press the button to hide the points.

6. Challenge your partner to match each measured slope with a line. Your partner must drag each measurement on top of the line it matches.

7. When your partner has finished guessing, press the button again to show the points. Your partner receives one point for each correctly matched slope.

You can scramble the lines automatically by selecting them and making an Animation button. Press the button to start scrambling the lines; press it again to stop.

8. Switch roles. While you look away, your partner will scramble the lines and hide the points. Now it's your turn to match the slopes with the lines.

9. After a round or two, you can add more lines to make the game more challenging.

Exploring Algebra 1 with The Geometer's Sketchpad

Distributing copies of the student notes is not really necessary. There are no questions to answer. The rules are quite simple, and they also appear on the first page of the Sketchpad document itself.

With all these games, have students work in pairs if possible. A sense of competition will tend to keep them on task.

GUESS THE SLOPE

The only way to lose points is to guess the sign of the slope incorrectly. This is perhaps the most common mistake. Students tend to equate slope with steepness, disregarding the direction of the pitch. Encourage students to make a decision about the sign of the slope at the outset.

Students should also build an inventory of mental reference images. With practice, it does not take long to visualize lines with slopes of 1, 0, or −1.

GUESS THE LINE

A good tool to employ here is the rise/run formula for slope. Form an imaginary right triangle using the segment between the control points as the hypotenuse.

MATCH THE SLOPE

Students benefit most by playing this game as two-person teams, so that team members can work together to formulate strategies for comparing line slopes.

When matching a slope measurement to a line, drag the measurement so it is centered directly over the line. You score a point only if the center of the measurement is close to the line without being closer to another line.

SLOPE ARCHERY

This is actually a variation of the Guess the Slope game, but with a higher degree of difficulty. Students must guess the slope of a line that is not even shown.

One predictable mistake will occur when the target is below and to the left of the archer. Students will tend to associate both of these directions with negative, but taken together, down and left produce a positive slope.

The Slope Archery game also involves an element of chance. The target's distance is variable because it is on the rectangular boundary of the archery range.

Exploring Algebra 1 with The Geometer's Sketchpad
© 2012 Key Curriculum Press

More Slope Games

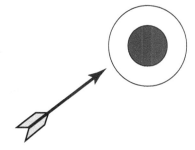

Nothing improves your skills like practice. Open **More Slope Games.gsp.** The document contains four games designed to improve your understanding of slope. You can set the games up for one player or two.

GUESS THE SLOPE

Start the game with the *Play* button. You will see a random line. Edit the parameter *Guess Slope* by double-clicking it, making your best guess for the slope of the line, and pressing OK. Press the *Enter* button to see your score. You get one point if your line is within 10° of the correct slope and two points if you are within 5°. Be careful to get the sign of the slope right. Get it wrong and you will lose one point, no matter how close your guess is.

GUESS THE LINE

This is the same game in reverse. When you press *Play,* you will see a slope. Try to make the line match that slope by dragging the points. Scoring is the same as in Guess the Slope.

MATCH THE SLOPE

This game gives you five lines and five slopes. Drag the slope measurements onto their corresponding lines. You must place the center of the measurement on (or very close to) the line to get credit. This game is best if you play it as part of a two-person team competing against another two-person team.

SLOPE ARCHERY

The red point is an archer's shooting position. When you press *Play,* the target will move to a new random location on the border of the range. Change the *Guess Slope* parameter to match an imaginary line between the archer and the target. Press *Enter* to shoot. A hit scores one or two points. There is no penalty for misses, no matter how bad.

How Slope Is Measured

Prior to starting this activity, you may wish to have a discussion on how students could measure slope or steepness—of a hill or staircase, for instance. The objective isn't to get students to discover the rule, but rather to help them appreciate the problem.

Discuss $\overleftrightarrow{AB}$ versus $\overline{AB}$. Discuss why the slope of $\overline{AB}$ represents the slope of the entire line. Ask, "Can you have a steep step on a staircase that isn't steep? Or vice versa?" (No, because the steepness of each step is the same as the steepness of the staircase. You may wish to point out the similar triangles.)

Be sure to tell students whether to collect their data on paper or in the Sketchpad table. If you have printers, it may be convenient for students to collect their data in the sketch and print the sketch when they finish the activity.

SKETCH AND INVESTIGATE

Q1 Here is the completed table (with answers in bold):

(x_A, y_A)	(x_B, y_B)	rise	run	Slope
(2, 1)	(4, 2)	1	2	0.5
(4, 0)	(5, 3)	**3**	**1**	**3**
(−5, −1)	(−3, 4)	**5**	**2**	**2.5**
(−5, 3)	(5, 4)	**1**	**10**	**0.1**
(2, −3)	**(4, 3)**	6	2	3

Q2 Here is the completed table (with answers in bold):

(x_A, y_A)	(x_B, y_B)	rise	run	Slope
(2, 1)	(4, 0)	**−1**	**2**	**−0.5**
(1, −1)	(0, 4)	**5**	**−1**	**−5**
(−3, 6)	(−5, −1)	**−7**	**−2**	**3.5**
(3, 5)	**(−1, 2)**	−3	−4	**0.75**

Q3 Switching A and B makes no difference, since the line and the step are still the same.

Q4 When B is above and to the left of A, the slope is negative; when B is below and to the left of A, the slope is positive. If *rise* and *run* have different signs, the slope is negative. If they have the same signs, the slope is positive.

Exploring Algebra 1 with The Geometer's Sketchpad
© 2012 Key Curriculum Press

Q5 Here are all possible integer answers in the original sketch window (with answers in bold):

(x_A, y_A)	(x_B, y_B)	rise	run	Slope
(1, 1)	(3, 2)	1	2	0.5
(1, 1)	**(5, 3)**	**2**	**4**	0.5
(1, 1)	**(7, 4)**	**3**	**6**	0.5
(1, 1)	**(−1, 0)**	**−1**	**−2**	0.5
(1, 1)	**(−3, −1)**	**−2**	**−4**	0.5
(1, 1)	**(−5, −2)**	**−3**	**−6**	0.5
(1, 1)	**(−7, −3)**	**−4**	**−8**	0.5

Q6 They are all on the same line. They can be reached by starting from *A* and repeatedly moving up 1 unit and right 2 units or down 1 unit and left 2 units.

Q7 *slope* = *rise/run*

Q8 $rise = y_B - y_A$

Q9 $run = x_B - x_A$

Q10 $slope = (y_B - y_A)/(x_B - x_A)$

EXPLORE MORE

Q11 The slope is the same in either direction. If you go the opposite way, the rise and the run will be the opposite of what they were before, and the ratio will be the same.

Q12 For a horizontal line, *rise* = 0, and *slope* = 0/*run*, which is 0 for any value of *run*. For a vertical line, *run* = 0 for any value of *rise*, and *slope* = *rise*/0, which is undefined.

WHOLE-CLASS PRESENTATION

Use the sketch **Slope Measurement.gsp** to show how slope changes as you change the line. The *Step* button gives students a picture of rise and run as a "step." Ask students to visualize the subsequent steps on this staircase, and use the *Staircase* button to show them.

To build a staircase, contractors first need to determine the *total rise* (height) and the *total run* (length) of the staircase. If the total rise is large compared to the total run, the stairs will be steep (and dangerous!). If the total rise is small compared to the

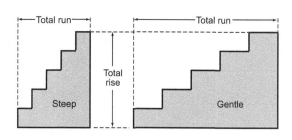

total run, the stairs will be easier to climb. So the relationship between the total rise and the total run determines the steepness of the staircase.

Each step also has a rise and run. The vertical part of a step is called the *riser,* and the horizontal part (where you step) is called the *tread.* A step with a large riser and small tread is steep. A step with a smaller riser and longer tread is safer and easier to climb. The steepness of each step depends on the overall steepness of the staircase. As you will see, this way of describing steepness is closely related to how slope is measured.

SKETCH AND INVESTIGATE

1. Open **Slope Measurement.gsp.** Press the *Show Coordinates* button.

Imagine building a staircase on this line, with one step going from point *A* to point *B*.

2. Press the *Step* button and observe the rise and the run. Drag *A* and *B*, and watch how these segments and values change.

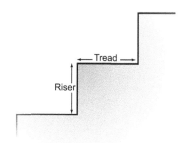

The rise is like the step riser and the run is like the tread. Press the *Show Staircase* button to see more stairs.

To keep track of your results using the table in the sketch, press the *Show Table* button. Double-click the table each time you want to add the current measurements to the table.

Q1 For each row of the table, drag *A* and *B* to match the given values, and fill in the rest of the row. (The first row has been filled in for you.)

A: (x_A, y_A)	B: (x_B, y_B)	*rise*	*run*	*Slope* $\overleftrightarrow{AB}$
(2, 1)	(4, 2)	1	2	0.5
(4, 0)	(5, 3)			
(−5, −1)	(−3, 4)			
(−5, 3)	(5, 4)			
(2, −3)	(,)	6	2	

Exploring Algebra 1 with The Geometer's Sketchpad
© 2012 Key Curriculum Press

Q2 In Q1, *B* was always above and to the right of *A*, and *rise* and *run* were always positive. What if *B* is below or to the left of *A*? Fill in the table to find out.

A: (x_A, y_A)	B: (x_B, y_B)	rise	run	Slope $\overleftrightarrow{AB}$
$(2, 1)$	$(4, 0)$			
$(1, -1)$	$(0, 4)$			
$(-3, 6)$	$(-5, -1)$			
$(3, 5)$	$(\quad , \quad)$	-3	-4	

Q3 What happens to *rise*, *run*, and *Slope* $\overleftrightarrow{AB}$ if you switch *A* and *B*? Try this for some of the table values above. Explain your results.

Q4 What happens to *Slope* $\overleftrightarrow{AB}$ when *B* is above and to the left of *A*? What happens when *B* is below and to the left of *A*? Why do you think this happens?

Q5 Fill in the following table with three other possible locations for point *B*.

If you have a printer and have kept your results in the Sketchpad table, you can use **File** | **Print** to print the sketch (including the table).

A: (x_A, y_A)	B: (x_B, y_B)	rise	run	Slope $\overleftrightarrow{AB}$
$(1, 1)$	$(3, 2)$	1	2	0.5
$(1, 1)$	$(\quad , \quad)$			0.5
$(1, 1)$	$(\quad , \quad)$			0.5
$(1, 1)$	$(\quad , \quad)$			0.5

Q6 Describe the locations for *B* that give a slope of 0.5. Explain why you think this happens. Move *A* to a different location. Does your explanation still work?

Q7 Looking back at your tables, you should notice a relationship between *rise*, *run*, and *Slope* $\overleftrightarrow{AB}$. Write a formula for *Slope* $\overleftrightarrow{AB}$ that uses *rise* and *run*.

Q8 Write a simple formula for *rise* that uses some or all of x_A, y_A, x_B, and y_B.

Q9 Write a simple formula for *run* that uses some or all of x_A, y_A, x_B, and y_B.

Q10 Rewrite your formula for *Slope* $\overleftrightarrow{AB}$ using x_A, y_A, x_B, and y_B.

EXPLORE MORE

Q11 So far, you've thought of *rise* as going up or down from point *A* and *run* as going right or left from there to point *B*. Would the slope be different if you went the other way? Press the *Show B to A* button. You'll see two new segments, *RISE* and *RUN*, going from *B* to *A*. Why is the slope the same whether you go from *A* to *B* along *rise* and *run* or from *B* to *A* along *RISE* and *RUN*?

Q12 In the activity The Slope of a Line, you learn that the slope of any horizontal line is 0 and the slope of any vertical line is undefined. Explain why this makes sense now that you know how slope is measured.

Slopes of Parallel and Perpendicular Lines  ACTIVITY NOTES

SKETCH AND INVESTIGATE

Q1 The angle measurement approaches 0° and should eventually disappear when the slopes are equal.

Q2 The vertex *E* of ∠*AEC* is the point of intersection of the lines. Lines with equal slope do not intersect, so the intersection point disappears when the slopes are the same; therefore, the angle also disappears.

Q3 If two lines have equal slopes, then the lines are parallel.

Q4 The product of the slopes of perpendicular lines is always −1 (as long as one of the lines is not horizontal).

Q5 If two lines are perpendicular, one of the slopes must be positive and the other negative. The product of a positive number and a negative number is always a negative number.

Q6 The slope of a vertical line is undefined, so the product of an undefined quantity with any other number is also undefined.

EXPLORE MORE

9. Select the line and choose **Equation** in the Measure menu. The coefficient of the *x* term in the equation equals the slope of the line (unless the line is vertical and has an undefined slope). For further development of this topic see the activity Different Slopes: The Slope of a Line.

10. Students should locate the second point by using the same rise and run as the two points on the first line.

11. If students construct parallel lines, the slopes will always be exactly equal. The product of the slopes of constructed perpendicular lines will always be exactly −1.

Slopes of Parallel and Perpendicular Lines

In this investigation you'll learn how you can use slope to tell whether lines are parallel or perpendicular.

SKETCH AND INVESTIGATE

Choose **Preferences** from the Edit menu and go to the Units panel.

1. In Preferences, set Angle Precision to **tenths** and precision of **Others** to **hundredths.**

2. Construct $\overleftrightarrow{AB}$ and $\overleftrightarrow{CD}$ and their point of intersection, *E*.

$m\angle AEC = 38.3°$

Slope $\overleftrightarrow{AB} = 0.56$

Slope $\overleftrightarrow{CD} = -0.16$

Select, in order, points *A*, *E*, and *C*; then, in the Measure menu, choose **Angle.**

3. Measure $\angle AEC$.

4. Measure the slopes of $\overleftrightarrow{AB}$ and $\overleftrightarrow{CD}$.

While holding down the Shift key, choose **Hide Coordinate System** from the Graph menu.

5. Hide the coordinate system that appeared when measuring the slopes.

6. Drag point *A* and observe the slope measures.

Q1 Make the slopes as close to equal as you can. What do you observe about the measure of the angle between the lines?

Q2 If you get the slopes close enough to equal, the angle measure will actually disappear. Why do you think that happens? (*Hint:* The vertex of this angle is the point of intersection of the two lines.)

Q3 Write a conjecture about lines with equal slopes.

Choose **Calculate** from the Number menu to open the Calculator. Click a measurement to enter it into a calculation.

7. Calculate the product of the slopes of $\overleftrightarrow{AB}$ and $\overleftrightarrow{CD}$.

8. Make sure that neither line is horizontal. Drag points to make m$\angle AEC$ as close to 90° as you can.

Q4 What is the product of the slopes of perpendicular lines? _____

Q5 Why is this product always negative?

Q6 The product of the slopes of two lines is undefined if one of the lines is vertical. Why?

EXPLORE MORE

9. In the same sketch, mesure the equations of the two lines. Where does the slope of a line appear in its equation?

It may help to choose
Snap Points from the
Graph menu.

10. In a new sketch, show the coordinate grid. Scale the grid, if necessary, so that grid points are about 1/2 in. (or 1 cm) apart. Hide the axes. Draw a line and a point not on the line. Now construct a second point not on the line, located so that when you draw a second line through these points it will be parallel to the first line. Explain how you located the second point.

11. Confirm your parallel-line slope conjecture by constructing a line and a point not on the line. Through the point not on the line, construct a parallel line. Measure the slopes of the two lines. Drag different points and observe the slope measurements. Do a similar investigation for perpendicular lines. Explain what you did and what your investigations demonstrate.

Exploring Algebra 1 with The Geometer's Sketchpad
© 2012 Key Curriculum Press

COMPARING SQUARES

Q1 $a = 2$, $b = 3$, and $a^2 + b^2 = 2^2 + 3^2 = 13$

Q2 The area of a square is the square of the length of a side.

Q3 The area of square $CFGH$ is $5^2 = 25$.

Q4 Each triangle has an area of 3. Their combined area is 12.

Q5 If you remove the four triangles from the circumscribed square, what remains is the hypotenuse square, having area c^2.

$$c^2 = 25 - 12 = 13$$

This matches $a^2 + b^2$ from Q1, so $a^2 + b^2 = c^2$. This supports the theorem.

Q6 When $a = 4$ and $b = 7$, $a^2 + b^2 = 65$.

The area of the circumscribed square is $11^2 = 121$.

Each triangle has area 14, so their combined area is 56.

The area of the hypotenuse square is $121 - 56 = 65$, so again, $a^2 + b^2 = c^2$.

THE DISTANCE FORMULA

Q7 The coordinates of point C are (x_B, y_A).

Q8 $a = x_B - x_A$ and $b = y_B - y_A$

Q9 When B is to the left of A, the calculation for side b is negative. It has the correct magnitude, but the sign is negative. In the distance formula, the calculation is squared. The end result will be correct because $(-b)^2 = b^2$. The same applies to the calculation for a.

Q10 The calculation will match the measurement no matter what the positions of A and B are. This is one math formula with no special cases or exceptions.

Q11 $d = \sqrt{(x_B - x_A)^2 + (y_B - y_A)^2}$

WHOLE-CLASS PRESENTATION

Comparing Squares

1. Open **Pythagorean Theorem Present.gsp.** Drag points A and B around, and let the class see that the triangle vertices always fall on grid intersections.

Q1 Ask someone to explain what the Pythagorean theorem means as applied to this triangle. [The legs of the triangle are *a* and *b*, and the hypotenuse is *c*, so $a^2 + b^2 = c^2$.] Press the *Show Pythagorean Theorem* button.

2. Use small numbers the first time. Drag the vertices into a position such that $a = 3$ and $b = 2$.

3. Press the *Show Leg Squares* and *Show Hypotenuse Square* buttons.

Q2 Ask what the square areas represent. [They represent the squares of the three triangle sides.]

Q3 Show that you can substitute $a^2 = 4$ and $b^2 = 9$ into the equation. Challenge the class to find the area of the hypotenuse square without using the Pythagorean theorem.

4. Press the *Show Circumscribed Square* button.

Q4 Ask for the area of the large square [25].

Q5 Ask for the areas of the four triangles. [Each of them has an area of 3.]

Q6 Now ask again for the area of the hypotenuse square. Students should see that you can get this by subtracting the four triangle areas from the circumscribed square area: $25 - 4(3) = 13$

5. Make this substitution in the equation, showing that the Pythagorean theorem works in this one case. Drag the vertex points and try at least one other case.

Distance Formula

6. Go to page 2. You will see points *A* and *B* on the coordinate grid. Tell the class that the objective is to calculate the coordinate distance between the points using only their coordinates.

7. Press the *Show Coordinates* button.

Q7 Press the *Show Triangle* button. Ask the class for the coordinates of point *C*. [The coordinates are (x_B, y_A).]

Q8 Since the legs of the right triangle are vertical and horizontal, it is a simple matter to find their lengths by subtracting coordinates. Ask for the formulas: $a = y_B - y_A$, $b = x_B - x_A$.

8. Press the *Show Calculations for a and b* button.

Exploring Algebra 1 with The Geometer's Sketchpad
© 2012 Key Curriculum Press

Q9 At this point, you have the lengths of the legs of the right triangle. You can use the Pythagorean theorem to compute distance *d*:

$$d = \sqrt{(x_B - x_A)^2 + (y_B - y_A)^2}$$

9. Press the *Show Calculation for d* button.

10. Select points *A* and *B*. Choose **Measure | Coordinate Distance.** Compare this distance to the one you calculated. You can explain to the class that this measurement is actually found using the same formula.

Q10 Drag points *A* and *B* around the screen to confirm that the calculation and the distance always concur. Depending on the positions of the points, either of the calculations for *a* and *b* can be negative. That cannot be right, since these numbers represent segment lengths. Ask for a good explanation of why the formula works even when these signs are wrong. [It is because the formula squares both of these calculations, making their signs irrelevant.]

The Pythagorean Theorem

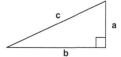

The Pythagorean theorem says that the sum of the squares of the lengths of the legs of a right triangle is equal to the square of the length of the hypotenuse. For the right triangle shown here, you can write the Pythagorean theorem as an equation.

$$a^2 + b^2 = c^2$$

To prove the theorem, you must show that it is true for all right triangles. In this activity you will demonstrate the theorem for several specific cases.

COMPARING SQUARES

1. Open **Pythagorean Theorem.gsp.** You see right triangle *ABC* on a square grid.

Don't use **Measure | Distance** or **Measure | Area** in this activity, because you do not know the scale of the grid.

Q1 How long are *a* and *b*? What is the sum of the squares of the lengths of the legs ($a^2 + b^2$)? Don't measure—just count grid squares.

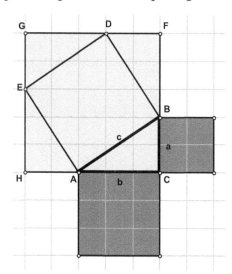

2. Press the *Show Leg Squares* button.

Q2 Explain why the areas of these squares are a^2 and b^2.

3. Press the *Show Hypotenuse Square* button.

This square, *ABDE*, has an area of c^2. Unfortunately, it is not aligned with the grid, so finding its area is more difficult.

4. Press the *Show Circumscribed Square* button.

Q3 This square, *CFGH*, fits around the hypotenuse square. What is its area?

Q4 There are four right triangles that also fit into the big square: △*ABC*, △*BDF*, △*DEG*, and △*EAH*. What is the area of each triangle? What is the sum of the triangle areas?

Q5 Use your answers to Q3 and Q4 to find the area of the hypotenuse square, *ABDE*. Does this support the Pythagorean theorem?

5. Drag the vertices of the triangle. You can change its dimensions, but it will always be a right triangle.

Q6 Change the leg dimensions to $a = 4$, $b = 7$. Using the same procedure as above, show that this triangle supports the Pythagorean theorem too.

Exploring Algebra 1 with The Geometer's Sketchpad
© 2012 Key Curriculum Press

THE DISTANCE FORMULA

In the steps below you will use the Pythagorean theorem to find the distance between two specific points from their coordinates. You will finish by writing a general formula for the distance between any two points on the coordinate plane.

6. Go to page 2. This page contains points *A* and *B* on the coordinate grid.

7. Press the *Show Triangle* button. This shows you right triangle *ABC*, like the triangle in the previous section but with the hypotenuse labeled *d* for "distance."

Triangle *ABC* is a right triangle, so the Pythagorean theorem should apply.

$$d^2 = a^2 + b^2$$
$$d = \sqrt{a^2 + b^2}$$

8. Select points *A* and *B*. Choose **Measure | Abscissae (x)**. Select *A* and *B* again, and choose **Measure | Ordinates (y)**.

Q7 You don't have to measure the coordinates of point *C* separately. What are the coordinates of C in terms of x_A, x_B, y_A, and y_B?

Q8 In terms of the coordinates, what are the lengths of sides *a* and *b*?

To enter a coordinate into the Calculator, click the coordinate measurement in the sketch.

9. Choose **Number | Calculate**. Calculate sides *a* and *b* from the coordinates.

10. Choose **Number | Calculate**. Using your calculations for *a* and *b*, use the Pythagorean theorem to calculate distance *d*.

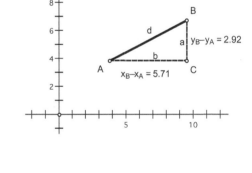

Q9 Drag point *B* so that it is left of *A*. What happens to the calculated distance *b*? Explain why this does not affect the final distance calculation.

11. Select *A* and *B*. Choose **Measure | Coordinate Distance**. Compare this distance to the one you calculated.

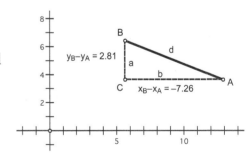

Q10 Drag points *A* and *B* to different locations on the coordinate plane. Does your calculated distance always match the measurement from step 11? Are there any special conditions under which the distance formula does not apply?

The label of your distance calculation should match the distance formula.

Q11 Write down a formula for the distance *d* in terms of the coordinates of *A* and *B*. This is the *distance formula*.

6

Variations and Linear Equations

Direct Variation

Students construct a rectangle model, take measurements, and then plot the point (*height, area*). They observe the dynamic trace of the plotted point as they change the height of the rectangle and keep the base constant. The result is a concrete and compelling demonstration of direct variation.

Inverse Variation

In the context of a time-distance problem, students explore an indirect variation in the form $xy = k$. They graph (x, y) points and then graph a family of curves.

The Slope-Intercept Form of a Line

Students explore the effects of m and b on the position of a line in the form $y = mx + b$. They write equations in slope-intercept form and visualize the graph when given the equation in slope-intercept form. (The activity Alternate Slope-Intercept Form of a Line is similar, but uses $y = a + bx$.)

Alternate Slope-Intercept Form of a Line

Students explore the effects of intercept and slope on the position of a line in the form $y = a + bx$. They practice writing equations in slope-intercept form, and visualizing the graph when given the equation in slope-intercept form.

The Point-Slope Form of a Line

Students use Sketchpad's dynamic capabilities to examine the effect of each constant on a linear equation written in point-slope form: $y = m(x - h) + k$. This activity helps to demystify the point-slope form. Note that the activity Alternate Point-Slope Form of a Line is similar but uses $y = y_1 + b(x - x_1)$.

Alternate Point-Slope Form of a Line

Students use Sketchpad's dynamic capabilities to examine the effect of each constant on a linear equation written in point-slope form: $y = y_1 + b(x - x_1)$. This activity helps to demystify the point-slope form. .

The Standard Form of a Line

Students explore the graph of a line expressed in the form $ax + by = c$, investigate the result of changing each of the values a, b, and c, and determine how these parameters relate to the slope and intercepts of the graph. They use their conclusions to write equations in standard form for several lines, given various pieces of information about the lines.

Lines of Fit

In the context of an archaeological find, students construct a scatter plot of eight data points. They approximate a line of best fit, and use that to make an estimate.

Direct Variation

In this activity students move from looking at properties of lines (in particular, slope) to generating linear relationships. Before starting, you might have a discussion about quantities that grow in different ways (proportionally, exponentially, and inversely) in relation to each other—and how each of these would look on a graph. The activity Inverse Variation focuses on quantities that are inversely proportional.

This activity focuses on lines of the form $y = mx$ (or $y = bx$), and another activity (The Slope-Intercept Form of a Line) focuses on lines of the form $y = mx + b$. It would be valuable to make the connection here between direct relationships and linear relationships (of which direct relationships are a subset).

SKETCH AND INVESTIGATE

1. If Sketchpad is set to its default Preference settings, points won't be labeled when they are created. Students can click points with the **Text** tool to label them. (Points will be labelled in alphabetical order.) To edit a label, double-click it with the **Text** tool.

Q1 Dragging B changes only the base and the area. Dragging C changes only the height and the area.

12. The sketch becomes cluttered at this point. Students may want to move the origin down near the bottom of the sketch window, hide the grid, and move the rectangle to a relatively clear area of the sketch.

Q2 It shows that as the height gets bigger, the area gets bigger; that as the height gets smaller, the area gets smaller; and that they grow or shrink proportionally to each other.

Q3 $A = base \cdot height$

Q4 $f(x) = base \cdot x$

Q5 It's the same. (To be more precise, it contains the path of the plotted point; the function exists in the first and fourth quadrants, but the plotted point is always in the first quadrant.)

Q6 The graph passes through the origin because the area of a rectangle with height 0 is 0—hence the point $(0, 0)$. And algebraically, when $x = 0$ in $f(x) = base \cdot x$, $f(0) = 0$—hence the point $(0, 0)$.

Q7 A rectangle can't have a negative height or area. The domain should be restricted to $x > 0$ (or possibly $x \geq 0$ if you consider 0 to be a possible height of a rectangle).

Q8 It means that as one quantity doubles, the other doubles; as one triples or halves, the other triples or halves. For example, the area of a rectangle with base 3 and height 4 is 12. If you double the height to 8 (and leave the base the same), the area also doubles to 24. The word "proportional" is used because the area and the height are in proportion ($12/4 = 24/8 = 3$). The base is the constant of proportionality.

Q9 This changes the slope of the line. Students may also notice that the plotted point moves vertically up and down (which makes sense because the x-value, which represents height, is not changing).

Q10 The length of the base is the slope of the graph. "Wide" rectangles (those with larger bases) will have steeper graphs. The reason is that every increase in height will add a lot to the area. "Skinny" rectangles (those with smaller bases) will have more gradual graphs. The reason is that similar increases in height will add much less to the area.

EXPLORE MORE

Q11 The point traces out a portion of a parabola. The trace is no longer linear, so this is *not* direct variation. The reason this happens is that we are now varying both the height and the base simultaneously, whereas before we were varying only the height, leaving the base constant. Variation in one dimension results in a linear graph, whereas variation in two dimensions results in a quadratic graph.

WHOLE-CLASS PRESENTATION

Use the sketch **Direct Variation Present.gsp** to explore with the class the relationships between measurements in a rectangle. The goal is to use this visual tool (and hopefully students' intuition) to think about why certain quantities are proportional and to connect this proportionality to direct variation via an equation and a graph.

Use page 1 of the sketch to answer Q1–Q10 together as a class. Use page 2 for Explore More (optional), which features a square instead of a rectangle. In this case, the area also increases when the side length increases, but the relationship is quadratic rather than linear.

Direct Variation

What happens to the area of a rectangle if you keep the length of the base constant while varying the height? (Try to answer this question before reading on.) What happens if you enlarge the entire rectangle? In this activity you will learn about direct variation and how it's represented algebraically and graphically.

SKETCH

Start by constructing a rectangle and its interior.

1. In a new sketch, construct a point and label it *A*.

To translate *A*, select it and choose **Transform | Translate.** To construct the ray, select the two points and choose **Construct | Ray.**

2. Translate point *A* by 1.0 cm at 0°. Construct a horizontal ray from *A* through the translated point.

3. Hide the translated point, construct a new point on the ray, and label it *B*.

4. Translate *B* by 1.0 cm at 90°. Construct a vertical ray from *B* through the translated point.

5. Hide the translated point, construct a new point on the ray, and label it *C*.

Select *B* and *A* in order and choose **Transform | Mark Vector.**

6. Mark the vector from point *B* to point *A*.

7. Translate point *C* by the marked vector. Label the translated point *D*.

8. Hide the rays and construct $\overline{AB}$, $\overline{BC}$, $\overline{CD}$, and $\overline{DA}$.

You should have a rectangle. Drag each of the four points to be sure it remains a rectangle and to see how dragging each point changes the rectangle.

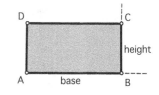

Select the four points in order and choose **Construct | Quadrilateral Interior.**

9. Construct polygon interior *ABCD*. Measure its area.

10. Measure the lengths of base $\overline{AB}$ and height $\overline{BC}$ by selecting them and choosing **Measure | Length.**

11. Click the **Text** tool on $\overline{AB}$ and $\overline{BC}$ to show their labels. Change $\overline{AB}$'s label to *base* and $\overline{BC}$'s label to *height* by double-clicking the **Text** tool on each label.

INVESTIGATE

Next you'll investigate how the area changes as you change the size of the rectangle.

Q1 Drag the points around. How do the measurements change when you drag *B*? How do the measurements change when you drag *C*?

If the sketch is too cluttered, choose **Graph | Hide Grid.**

12. A graph will help to show what's happening. Select in order the *height* and *Area ABCD* measurements, and choose **Graph | Plot as (x, y).**

You have just plotted a point whose *x*- and *y*-coordinates are the height and area of the rectangle, respectively. (If you can't see it yet, you will soon.)

To trace an object, select it and choose **Display | Trace Plotted Point.** To clear traces from the screen, choose **Display | Erase Traces.**

13. Drag point *C* closer to $\overline{AB}$ and observe the effect on the plotted point. Trace the plotted point so you can see what it does as you drag point *C*.

Q2 What does the path of the plotted point tell you about how height and area are related in a rectangle when the base is kept constant?

Q3 Write the formula for the area of a rectangle in terms of *base* and *height*.

Q4 Now write the same formula as a function using $f(x)$ for *area* and *x* for *height*.

To enter *base* in your formula, click its measurement in the sketch. To enter *x*, click the *x* key on the Calculator's keypad.

Q5 Plot your function from Q4 by choosing **Graph | Plot New Function.** How does the function plot relate to the path of the plotted point as you vary the height of the rectangle?

Q6 Why does it make sense that the graph passes through the origin?

Q7 The function is plotted as a line. But the traces cover only part of the line. Why? If you wanted the function plot to accurately represent the situation, what part of it would you cut off? (In other words, how would you restrict the *domain*?)

Q8 We say that a rectangle's area *varies directly with* (or *is directly proportional to*) its height when its base is held constant. Describe in your own words what you think this means.

To answer Q10, compare a rectangle that produces a steep graph to one that produces a gradual graph.

Q9 Erase the traces. Drag point *B* to change the length of the base of the rectangle. What effect does this have on the graph?

Q10 How is the length of the base related to the slope of the graph?

EXPLORE MORE

Q11 Go to page 2 of **Direct Variation.gsp.** This page shows square *ABCD*. Does the area of the square vary directly with the length of one side? Make a prediction and then drag the points to check your prediction. Describe the path of the plotted point. Is this direct variation? Why or why not?

Exploring Algebra 1 with The Geometer's Sketchpad
© 2012 Key Curriculum Press

Inverse Variation

PUNCTUALITY

Q1 Yes, this is an inverse variation. Speed and time are variables, and the distance is a constant. In this context, distance is the distance from Adrianne's house to her school. That distance cannot change, so it is reasonable to call it a constant.

Q2 At 6:30, the time remaining (x) is 1.5 hours. The distance is 3 miles. Solving for y, Adrianne's average speed must be 2 mi/h.

$x = 1.5, y = 2$

Q3 She must walk 3 mi/h.

$x = 1, y = 3$

2. Check to see that everyone is plotting the first two points correctly, $(1.5, 2)$ and $(1, 3)$. After that, students should have no trouble plotting points on their own.

Q4 She has 0.75 hour remaining. She must travel 4 mi/h.

$(0.75, 4)$

Q5 She will have 0.5 hour remaining. She must ride her bicycle at 6 mi/h.

$(0.5, 6)$

Q6 She will have 1/3 hour remaining. The bus must travel 9 mi/h.

$(1/3, 9)$

It is possible to enter the fraction 1/3 in the Plot Points dialog box, or students can enter a decimal approximation.

Q7 There will be 1/6 hour remaining. Pete must drive 18 mi/h.

$(1/6, 18)$

Eighteen mi/h may sound slow for a car. Remind students that this is the average speed. Pete has to allow for traffic lights and stop signs, and he will have to find a parking space.

Q8 Answers will vary. Some students will observe that as one of the variables increases, the other decreases. Other students may be more precise, pointing out that the required speed is proportional to the multiplicative inverse of the remaining time.

GRAPH THE CURVE

Q9 $y = \frac{3}{x}$

Students may have already written the equation in this form. It would make it easier to answer the previous questions.

Q10 Since the curve represents the general solution for any given x, all of the plotted points should lie on the curve.

Q11 It is not possible for any two of the curves to intersect. If they did, the intersection point (x, y) would satisfy both equations. That means that the time and speed would be the same in both cases. If the time and speed are the same, then the distance traveled must be the same, but each curve represents a different travel distance.

This activity investigates the properties of an indirect variation in this form:

$$xy = k, \text{ where } k \text{ is a constant}$$

Tell students to imagine that they live 3 miles from school and that they must arrive there by 8:00. If they know how much time they have when they leave the house, they can compute what their average speed must be in order to arrive on time.

1. Write this formula: *speed · time = distance*

Q1 Ask students which number is a constant. It must be the distance because the distance between home and school does not change.

2. Define variables and units, and write them on the board. Write the equation too.

$$x = \text{time remaining before 8:00 (hours)}$$

$$y = \text{speed (mi/h)}$$

$$xy = 3$$

3. Draw the following table, with only the top row filled in:

	6:30	7:00	7:15	7:30	7:40	7:50
x	1.5	1	0.75	0.5	0.33	0.17
y	2	3	4	6	9	18

4. Model the first column for them. It is 6:30, so she has 1.5 hours to get to class. Substitute 1.5 for x in the equation, and solve for y.

5. Have students fill in the rest of the table.

Tell them to stay with the given units—no minutes or seconds.

Q2 Ask students to explain why this relationship is called an inverse variation. (As x becomes smaller, y grows larger, and vice versa.)

6. Open **Inverse Variation Present.gsp.** There is a set of coordinate axes at an appropriate scale. Ignore the buttons for now.

7. Choose **Graph | Plot Points.** Enter the (x, y) coordinates from the table.

Q3 What is the equation for y as a function of x? $\left[y = \frac{3}{x} \right]$

8. Press the *Show Graph* button to reveal the graph of the equation. Confirm that all of the plotted points are on the curve.

Q4 Challenge students to describe what the general shape of the graph would be if the distance were 1 mile. (Same, but closer to the axes.) What if it were 5 miles?

9. Press the *Show Other Curves* button to see the graph for 1, 2, 4, and 5 miles.

Q5 Is it possible for any of these graphs to intersect? Discuss.

Inverse Variation

Two variables, x and y, have an *inverse relationship* if y depends on the inverse of x:

$$y = \frac{k}{x}, \text{ where } k \text{ is a constant}$$

You can also express this in the form $xy = k$: The product of x and y is constant.

PUNCTUALITY

Adrianne has been late for school twice this week. She has resolved to become more punctual (and avoid detention). Her home is 3 miles from school, and the first class starts at 8:00. She figures that if she leaves the house at 6:30, she can walk to school at a leisurely pace and still arrive well before the bell. Before leaving, Adrianne does a quick calculation to see how fast she needs to walk. For this, she uses a formula that relates speed, time, and distance:

$$speed \cdot time = distance$$

Q1 Is this an inverse variation? Which numbers are the variables, and which is the constant? Is it reasonable to call that number a constant?

> Let the *x* units be hours, and let the *y* units be mi/h. Do not use any other units.

Q2 Let x be the time (in hours) remaining before class, and let y be the speed that Adrianne needs to travel in order to arrive at 8:00. It is now 6:30. What is x and what is y? How fast must Adrianne walk?

Q3 While Adrianne was looking up the formula, 30 minutes passed. It is now 7:00. Compute x and y again. How fast must she walk if she leaves now?

It will have to be a fast walk. She decides to graph it, just to be thorough.

1. In a new sketch, choose **Graph | Grid Form | Rectangular Grid.** The rectangular grid allows you to adjust the x and y scales independently.

2. Choose **Graph | Plot Points.** Plot the (x, y) pairs that you computed in Q2 and Q3.

> In the Plot Points dialog box, you can enter simple expressions. For example, enter 1/3 for one third of an hour (20 minutes).

Q4 The graph took longer than she expected. Now it's 7:15. How much time does she have left? How fast must she travel? Plot the point.

Q5 Now Adrianne has decided that it would be a better idea to ride her bicycle. There is still plenty of time if she leaves now (OK, as soon as this TV show is over). How fast will she have to ride if she leaves at 7:30? Plot the point.

Q6 Wouldn't you know it? Now it's raining. She is absolutely not going to school with wet hair. The bus will come by at about 7:40. How fast will it have to travel to get Adrianne to school on time? Plot the point.

Q7 Adrianne must have dozed. She's missed the bus. Now she will have to ride to school with her brother, Pete. He always drives too fast. He doesn't even leave the house until 7:50. How fast will he have to drive? Plot the point.

Q8 Look at the values you have found so far. In your own words, explain why the relationship between these two quantities is called *inverse variation*.

It's 7:50 now. Where's Pete? Adrianne's mother tells her that Pete took the bus. His car broke down yesterday. This is just too much! Now she'll have three tardy notices in one week, and it wasn't even her fault.

GRAPH THE CURVE

Adrianne now realizes that she was wasting too much time on calculations. Rather than compute each speed separately, she should have written *y* as a function of *x*. She could then plot the curve and find the speed for a whole range of departure times.

Q9 Write an expression for *y* in terms of *x*.

3. Choose **Graph | Plot New Function.** For the function definition, enter the formula for *y* from your answer to Q9.

Q10 What is the relationship between this curve and the points you plotted earlier?

To change the domain of a graph, select the graph and choose **Edit | Properties | Plot.** Negative time would make no sense here, so let zero be the minimum value of *x*.

4. This graph applies only to students who live 3 miles from school. Draw similar graphs for students living 1, 2, 4, and 5 miles from school. Use different colors. Label the axes and the curves.

Q11 Do any of these curves intersect? Explain why or why not.

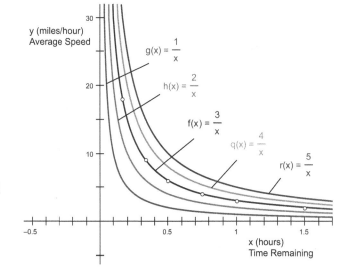

The value *m* is the slope of the line, and *b* is where the line crosses the *y*-axis. (This formula can also be written $y = a + bx$, using *a* and *b* instead of *b* and *m*. Some students may be more familiar with this form.)

SKETCH AND INVESTIGATE

1. Hiding the unit point (0, 1) reduces the chance that students will change the scale of the coordinate system. Sketchpad measures coordinates in graph units but does translation in distance units (usually cm). When the coordinate system is defined, those units agree. If the points in Q3 and Q4 do not have integer values, the student has probably changed the scale by dragging the unit point or the tick numbers on the axes.

Q1 When $x = 0$, $y = 1$. The point is (0, 1). It makes sense to call this the *y*-intercept because it's the point where the line crosses the *y*-axis.

Q2 The *y*-intercept of $y = 3x + 7$ is 7. When you substitute 0 for *x* in $y = mx + b$, you get $y = m(0) + b$, or $y = b$.

Q3 The coordinates of the new point are (1, 3). This satisfies the equation because $y = 2(1) + 1 = 3$. (See the note for step 1 if students get non-integer coordinates when they measure them.)

Q4 The third point is (2, 5). This point satisfies the equation because $y = 2(2) + 1 = 5$.

Q5 The lines are shown below with several integer points plotted.

a.
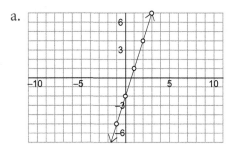

b.
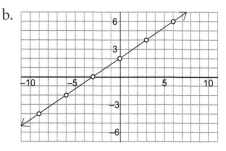

c.
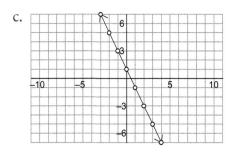

d.
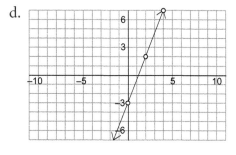

Exploring Algebra 1 with The Geometer's Sketchpad
© 2012 Key Curriculum Press

Q6 Lines with a positive *m* go up to the right and down to the left, lines with a negative *m* go down to the right and up to the left, and lines with $m = 0$ are horizontal. As *m* becomes increasingly positive or negative, the line becomes steeper.

Q7 As *b* becomes increasingly positive, the line is shifted (translated) up. As *b* becomes increasingly negative, the line is shifted (translated) down. When $b = 0$, the line goes through the origin.

Q8 The slopes vary, but the traces always pass through the same *y*-intercept. The result looks like an "infinite asterisk."

Q9 This family can be pictured as the infinite set of lines in a plane that are parallel to a given line. They all have the same slope.

Q10 a. $y = 2x - 3$ b. $y = -1.5x + 4$
       c. $y = 3x + 6$ d. $y = -0.4x - 0.4$
       e. $y = 0.5x + 3.5$

EXPLORE MORE

Q11 This line is parallel to the *y*-axis, so it has no *y*-intercept and the slope is undefined. The line can be expressed with the equation $x = 3$, but that's not in slope-intercept form.

Q12 No, it's not possible. The reason is that every line has a unique *y*-intercept, so there's only one value for *b* for a particular line. Similarly, each line has a unique slope, so there's only one value for *m*.

WHOLE-CLASS PRESENTATION

Use the sketch **Slope Intercept Present.gsp** to help students visualize the graph of a line from an equation written in slope-intercept form. You will need to discuss how the *y*-intercept is found by substituting 0 for *x*, which always yields $y = b$ for an equation in the form $y = mx + b$. Then the slope can be applied to find one or two more points and graph the line.

Use page 2 to further explore the effects of *m* and *b*. This sketch is set up with sliders for *m* and *b*. You can use this sketch to explore Q6–Q12 with the whole class.

The Slope-Intercept Form of a Line

The slope-intercept form of a line, $y = mx + b$, is one of the best-known formulas in algebra. In this activity you'll learn about this equation first by exploring one line, and then by exploring whole *families* of lines.

SKETCH AND INVESTIGATE

Choose **Graph | Define Coordinate System.** To hide the points, select them and choose **Display | Hide Points.**

You'll start this activity with $m = 2$ and $b = 1$ as you explore the line $y = 2x + 1$.

1. In a new sketch, define a coordinate system and hide the points $(0, 0)$ and $(1, 0)$.

Q1 For $y = 2x + 1$, what is y when $x = 0$? Write your answer as an ordered pair.

Choose **Graph | Plot Points.** Enter the coordinates in the Plot Points dialog box, click **Plot;** then click **Done.**

2. Plot this point. Why does it make sense to call this point the *y-intercept*?

Q2 You found that the y-intercept of $y = 2x + 1$ is 1. What is the y-intercept of $y = 3x + 7$? Explain why the y-intercept of $y = mx + b$ is always b.

You've learned that *slope* can be written as *rise/run*. The slope of the line $y = 2x + 1$ is 2, which you can think of as 2/1 (*rise* = 2 and *run* = 1).

3. Translate your plotted point using this slope. Choose **Transform | Translate,** use a rectangular translation vector, and enter 1 for the run (horizontal) and 2 for the rise (vertical).

To measure the coordinates, choose **Measure | Coordinates.**

Q3 What are the coordinates of the new point? Substitute them into $y = 2x + 1$ to show they satisfy the equation.

Q4 Translate the new point by the same *rise* and *run* values to get a third point. Find the coordinates of this third point, and verify that it satisfies the equation $y = 2x + 1$.

4. Select any two of the three points you've plotted, and choose **Construct | Line.**

What you've done so far is one technique for plotting lines in the form $y = mx + b$:

• Plot the y-intercept $(0, b)$.

If *m* is a decimal such as 1.5, write it as a fraction such as 3/2. If it's a whole number such as 3, write it as a fraction such as 3/1.

• Rewrite m as *rise/run* (if necessary).

• Find a second point by translating the y-intercept by *rise* and *run*.

• Connect the points to get the line. Plot a third point to check the line.

Q5 Using the method just described, plot these lines on graph paper.

    a. $y = 3x - 2$ b. $y = (2/3)x + 2$

    c. $y = -2x + 1$ d. $y = 2.5x - 3$

EXPLORING FAMILIES OF LINES

Now that you've plotted a line, focus on how m and b affect the equation.

5. Open **Slope Intercept.gsp.**

The graph of $y = 2x + 1$ is already plotted. You can change m and b by adjusting their sliders.

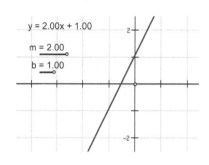

To adjust a slider, drag the point at its tip.

Q6 Adjust slider m and observe the effect. Describe the differences between lines with $m > 0$, $m < 0$, and $m = 0$. What happens to the line as m becomes increasingly positive? Increasingly negative?

Q7 Now adjust slider b. Describe the effect this value has on the line.

6. Select the line and choose **Display | Trace Line.**

Q8 Adjust m and observe the trace pattern that forms. Describe the lines that appear when you change m. What do they have in common?

To erase traces left by the line, choose **Display | Erase Traces.**

Q9 Erase the traces and adjust b. How would you describe the lines that form when you change b? What do they have in common?

7. Turn off tracing by selecting the line and choosing **Display | Trace Line** again. Erase any remaining traces.

Q10 For each description below, write the equation in slope-intercept form. To check your equation, adjust m and b so that the line appears on the screen.

    a. slope is 2.0; y-intercept is $(0, -3)$

    b. slope is -1.5; y-intercept is $(0, 4)$

    c. slope is 3.0; x-intercept is $(-2, 0)$

    d. slope is -0.4; contains the point $(-6, 2)$

    e. contains the points $(3, 5)$ and $(-1, 3)$

EXPLORE MORE

Q11 Attempt to construct a line through the points $(3, 0)$ and $(3, -3)$ by adjusting the sliders in the sketch. Explain why this is impossible. (Why can't you write its equation in slope-intercept form?)

Q12 Can you construct the same line with two different slider configurations? If so, provide two different equations for the same line. If not, explain why.

Alternate Slope-Intercept Form of a Line

The value *b* is the slope of the line, and *a* is where the line crosses the *y*-axis. (This formula can also be written $y = mx + b$, using *m* and *b* instead of *b* and *a*. Some students may be more familiar with this form.)

SKETCH AND INVESTIGATE

1. Hiding the unit point (0, 1) reduces the chance that students will change the scale of the coordinate system. Sketchpad measures coordinates in graph units but does translation in distance units (usually cm). When the coordinate system is defined, those units agree. If the points in Q3 and Q4 do not have integer values, the student has probably changed the scale by dragging the unit point or the tick numbers on the axes.

Q1 When $x = 0$, $y = 1$. The point is (0, 1). It makes sense to call this the *y*-intercept because it's the point where the line crosses the *y*-axis.

Q2 The *y*-intercept of $y = 7 + 3x$ is 7. When you substitute 0 for *x* in $y = a + bx$, you get $y = a + b(0)$, or $y = a$.

Q3 The coordinates of the new point are (1, 3). This satisfies the equation because $y = 1 + 2(1) = 3$. (See the note for step 1 if students get non-integer coordinates when they measure them.)

Q4 The third point is (2, 5). This point satisfies the equation because $y = 1 + 2(2) = 5$.

Q5 The lines are shown below with several integer points plotted.

a.
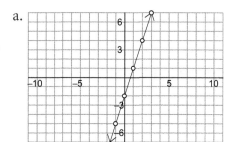

b.
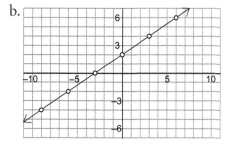

c.
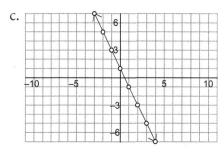

d.
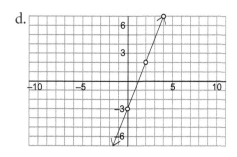

Q6 Lines with a positive b go up to the right and down to the left, lines with a negative b go down to the right and up to the left, and lines with $b = 0$ are horizontal. As b becomes increasingly positive or negative, the line becomes steeper.

Q7 As a becomes increasingly positive, the line is shifted (translated) up. As a becomes increasingly negative, the line is shifted (translated) down. When $a = 0$, the line goes through the origin.

Q8 The slopes vary, but the traces always pass through the same y-intercept. The result looks like an "infinite asterisk."

Q9 This family can be pictured as the infinite set of lines in a plane that are parallel to a given line. They all have the same slope.

Q10 a. $y = -3 + 2x$ b. $y = 4 - 1.5x$ c. $y = 6 + 3x$

d. $y = -0.4 - 0.4x$ e. $y = 3.5 + 0.5x$

EXPLORE MORE

Q11 This line is parallel to the y-axis, so it has no y-intercept and the slope is undefined. The line can be expressed with the equation $x = 3$, but that's not in slope-intercept form.

Q12 No, it's not possible. The reason is that every line has a unique y-intercept, so there's only one value for a for a particular line. Similarly, each line has a unique slope, so there's only one value for b.

WHOLE-CLASS PRESENTATION

Use the sketch **Slope Intercept2 Present.gsp** to help students visualize the graph of a line from an equation written in slope-intercept form. You will need to discuss how the y-intercept is found by substituting 0 for x, which always yields $y = a$ for an equation in the form $y = a + bx$. Then the slope can be applied to find one or two more points and graph the line.

Use page 2 to further explore the effects of a and b. This sketch is set up with sliders for a and b. You can use this sketch to explore Q6–Q12 with the whole class.

Alternate Slope-Intercept Form of a Line

The slope-intercept form of a line, $y = a + bx$, is one of the best-known formulas in algebra. In this activity you'll learn about this equation first by exploring one line, and then by exploring whole *families* of lines.

SKETCH AND INVESTIGATE

Choose **Graph | Define Coordinate System.** To hide the points, select them and choose **Display | Hide Points.**

You'll start this activity with $a = 1$ and $b = 2$ as you explore the line $y = 1 + 2x$.

1. In a new sketch, define a coordinate system and hide the points $(0, 0)$ and $(1, 0)$.

Q1 For $y = 1 + 2x$, what is y when $x = 0$? Write your answer as an ordered pair.

Choose **Graph | Plot Points.** Enter the coordinates in the Plot Points dialog box, click **Plot;** then click **Done.**

2. Plot this point. Why does it make sense to call this point the *y-intercept*?

Q2 You found that the *y*-intercept of $y = 1 + 2x$ is 1. What is the *y*-intercept of $y = 7 + 3x$? Explain why the *y*-intercept of $y = a + bx$ is always a.

You've learned that *slope* can be written as *rise/run*. The slope of the line $y = 1 + 2x$ is 2, which you can think of as 2/1 (*rise* = 2 and *run* = 1).

3. Translate your plotted point using this slope. Choose **Transform | Translate,** use a rectangular translation vector, and enter 1 for the run (horizontal) and 2 for the rise (vertical).

To measure the coordinates, choose **Measure | Coordinates.**

Q3 What are the coordinates of the new point? Substitute them into $y = 1 + 2x$ to show they satisfy the equation.

Q4 Translate the new point by the same *rise* and *run* values to get a third point. Find the coordinates of this third point, and verify that it satisfies the equation $y = 1 + 2x$.

4. Select any two of the three points you've plotted, and choose **Construct | Line.**

What you've done so far is one technique for plotting lines in the form $y = a + bx$:

- Plot the *y*-intercept $(0, a)$.

If *b* is a decimal such as 1.5, write it as a fraction such as 3/2. If it's a whole number such as 3, write it as a fraction such as 3/1.

- Rewrite b as *rise/run* (if necessary).

- Find a second point by translating the *y*-intercept by *rise* and *run*.

- Connect the points to get the line. Plot a third point to check the line.

Q5 Using the method just described, plot these lines on graph paper.

    a. $y = -2 + 3x$ b. $y = 2 + (2/3)x$

    c. $y = 1 - 2x$ d. $y = -3 + 2.5x$

EXPLORING FAMILIES OF LINES

Now that you've plotted a line, focus on how m and b affect the equation.

5. Open **Slope Intercept 2.gsp.**

The graph of $y = 1 + 2x$ is already plotted.
You can change a and b by adjusting their sliders.

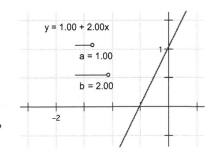

To adjust a slider, drag the point at its tip.

Q6 Adjust slider b and observe the effect. Describe the differences between lines with $b > 0$, $b < 0$, and $b = 0$. What happens to the line as b becomes increasingly positive? Increasingly negative?

Q7 Now adjust slider a. Describe the effect this value has on the line.

6. Select the line and choose **Display | Trace Line.**

Q8 Adjust b and observe the trace pattern that forms. Describe the lines that appear when you change b. What do they have in common?

To erase traces left by the line, choose **Display | Erase Traces.**

Q9 Erase the traces and adjust a. How would you describe the lines that form when you change a? What do they have in common?

7. Turn off tracing by selecting the line and choosing **Display | Trace Line** again. Erase any remaining traces.

Q10 For each description below, write the equation in slope-intercept form. To check your equation, adjust a and b so that the line appears on the screen.

    a. slope is 2.0; y-intercept is $(0, -3)$

    b. slope is -1.5; y-intercept is $(0, 4)$

    c. slope is 3.0; x-intercept is $(-2, 0)$

    d. slope is -0.4; contains the point $(-6, 2)$

    e. contains the points $(3, 5)$ and $(-1, 3)$

EXPLORE MORE

Q11 Attempt to construct a line through the points $(3, 0)$ and $(3, -3)$ by adjusting the sliders in the sketch. Explain why this is impossible. (Why can't you write its equation in slope-intercept form?)

Q12 Can you construct the same line with two different slider configurations? If so, provide two different equations for the same line. If not, explain why.

The Point-Slope Form of a Line

SKETCH AND INVESTIGATE

3. Students may have to drag the New Function dialog box by its title bar in order to see measurements *m*, *h*, and *k* in the sketch. Students will need to enter the implied multiplication sign after *m*.

Q1 The line appears to spin around the point (h, k).

Q2 This does support the answer from Q1. You can now see the center point as the line spins.

Q3 This is the family of lines through the point (h, k) with any slope. This family can be pictured as an asterisk with the center point (h, k).

Q4 These are families of lines with the same slope. The families can be pictured as the infinite set of lines in a plane parallel to a given line. It's interesting that although the two families look the same, they are formed in different ways (as you can see by watching the point (h, k). Adjusting *h* moves the lines right and left whereas adjusting *k* moves the lines up and down.

Q5 Parameter *h* is the *x*-coordinate of the special point that the line spins around when *m* is dragged. Making *h* larger moves the line to the right; making it smaller (or more negative) moves it to the left.

Parameter *k* is the *y*-coordinate of the special point that the line spins around when *m* is dragged. Making *k* larger moves the line up; making it smaller (more negative) moves it down.

Q6 $y = 2(x - 1) + 3$

Q7 a. $y = 2(x - (-2)) + 1$ or $y = 2(x + 2) + 1$
   b. $y = -1(x - (-2)) + 1$ or $y = -1(x + 2) + 1$
   c. $y = 3(x - 0) + 0$ or $y = 3x$
   d. $y = 0.8(x - 2) + 0$ or $y = 0.8(x - 2)$
   e. $y = (-1/3)(x - 2) + 3$ or $y = (-1/3)(x + 1) + 4$
   f. $y = 0(x - 4) + 5$ or $y = 5$

EXPLORE MORE

Q8 Slope is defined as *rise/run*. For this line, since the *x*-coordinates of both points are the same (2), *run* = 0. Thus the slope is undefined, since you can't divide by 0. The line can be expressed with the equation $x = 2$, but that's not in point-slope form.

Exploring Algebra 1 with The Geometer's Sketchpad
© 2012 Key Curriculum Press

Q9 It is possible to express any line (except vertical lines) with an infinite number of equations in point-slope form (or an infinite number of slider configurations in the sketch). The reason is that h and k aren't unique—any point on the line will do. For example, the line $y = 2(x - 1) + 3$ also goes through the points $(2, 5)$ and $(3, 7)$, so the equations $y = 2(x - 2) + 5$ and $y = 2(x - 3) + 7$ also express this same line. (Try it out!)

WHOLE-CLASS PRESENTATION

Students see the effects of changing each of the values m, h, and k. They see how changing these values in a linear equation in point-slope form changes the location of the line on the graph. Use the sketch **Point Slope Present.gsp.** Press the button *Show Point on Line*, and drag x to show how the calculated y value changes for different values of x. Then press the button *Show Function and Line*, and drag x again. Use the sliders to change m, h, and k, and let students make observations. Do Q1–Q7 together as a class.

The Point-Slope Form of a Line

The slope-intercept form of a line is great if you know one special point: the y-intercept. But what if the point you know is an everyday, ordinary point such as $(3, -2)$ or $(-7, -7)$? In this case it's usually most convenient to use the *point-slope form* of a line, which you'll study in this activity.

SKETCH AND INVESTIGATE

1. Open the sketch **Point Slope.gsp.**

To adjust a slider, drag the point at its tip.

You'll see an equation in the point-slope form $y = m(x - h) + k$, with numbers filled in for m, h, and k. Adjust the sliders for m, h, and k, and watch the equation change. There's no line yet, but you can graph one.

2. Choose **Graph | Plot New Function.**

The New Function dialog box appears.

To enter m, h, and k, click their measurements in the sketch. To enter x, click the x key on the keypad.

3. Enter $m(x - h) + k$ and click **OK.**

Sketchpad plots the function for the current values of the parameters m, h, and k.

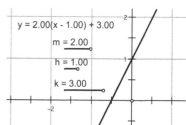

4. Select the new line and choose **Display | Trace Function Plot.**

If at any point you wish to erase traces left by the line, choose **Display | Erase Traces.**

Q1 Adjust slider m. You'll see that the line rotates around a single point. Change the values of h and k, then adjust m again, focusing on where this point appears to be. What are the point's coordinates? How do they relate to h and k?

To deselect all objects, click the **Arrow** tool in empty space.

5. Deselect all objects. Now select, in order, measurement h and measurement k. Choose **Graph | Plot as (x, y)** to plot the point (h, k).

Q2 Adjust slider m again and observe what happens. Does this support your answer from Q1?

Q3 Describe the family of lines that forms when you change m.

Q4 Adjust sliders h and k, one at a time. How would you describe the families of lines that form when varying each of these values? How do they compare to each other?

Q5 Summarize the roles that h and k play in the equation $y = m(x - h) + k$.

Q6 Suppose you know that the slope of a line is 2 and that it contains the point $(1, 3)$. What is the equation in point-slope form for this line? Check your answer by adjusting the sliders in the sketch.

Q7 Write an equation in point-slope form for each of the lines described. When you finish, check each of your answers by adjusting sliders m, h, and k so that the line is drawn on the screen.

    a. slope is 2; contains the point $(-2, 1)$

    b. slope is -1; contains the point $(-2, 1)$

    c. is parallel to the line $y = 3(x - 2) + 4$; goes through the origin

    d. slope is $\frac{4}{5}$; x-intercept is $(2, 0)$

    e. contains the points $(2, 3)$ and $(-1, 4)$

    f. contains the points $(-3, 5)$ and $(4, 5)$

EXPLORE MORE

Q8 Try to construct a line through the points $(2, 3)$ and $(2, -2)$ by adjusting the sliders in the sketch. Explain why this is impossible and why this equation cannot be written in point-slope form.

Q9 Is it possible to construct the same line with different slider configurations? If not, explain why. If so, provide two different equations for the same line.

Alternate Point-Slope Form of a Line

SKETCH AND INVESTIGATE

3. Students may have to drag the New Function dialog box by its title bar in order to see measurements y_1, b, and x_1 in the sketch. Students will need to enter the (implied) multiplication sign after b.

Q1 The line appears to spin around the point (x_1, y_1).

Q2 This does support the answer from Q1. You can now see the center point as the line spins.

Q3 This is the family of lines through the point (x_1, y_1) with any slope. This family can be pictured as an asterisk with the center point (x_1, y_1).

Q4 These are families of lines with the same slope. The families can be pictured as the infinite set of lines in a plane parallel to a given line. It's interesting that although the two families look the same, they are formed in different ways, as you can see by watching the point (x_1, y_1). Adjusting x_1 moves the lines right and left whereas adjusting y_1 moves the lines up and down.

Q5 Parameter x_1 is the x-coordinate of the special point that the line spins around when b is dragged. Making x_1 larger moves the line to the right; making it smaller (or more negative) moves it to the left.

Parameter y_1 is the y-coordinate of the special point that the line spins around when b is dragged. Making y_1 larger moves the line up; making it smaller (more negative) moves it down.

Q6 $y = 3 + 2(x - 1)$

Q7 a. $y = 1 + 2(x - (-2))$ or $y = 1 + 2(x + 2)$

b. $y = 1 - 1(x - (-2))$ or $y = 1 - (x + 2)$

c. $y = 0 + 3(x - 0)$ or $y = 3x$

d. $y = 0 + 0.8(x - 2)$ or $y = 0.8(x - 2)$

e. $y = 3 - (1/3)(x - 2)$ or $y = 4 - (1/3)(x + 1)$

f. $y = 5 + 0(x - 4)$ or $y = 5$

EXPLORE MORE

Q8 Slope is defined as *rise/run*. For this line, since the *x*-coordinates of both points are the same (2), *run* = 0. Thus the slope is undefined, since you can't divide by 0. The line can be expressed with the equation $x = 2$, but that's not in point-slope form.

Q9 It is possible to express any line (except vertical lines) with an infinite number of equations in point-slope form (or an infinite number of slider configurations in the sketch). The reason is that x_1 and y_1 aren't unique—any point on the line will do. For example, the line $y = 3 + 2(x - 1)$ also goes through the points (2, 5) and (3, 7), so the equations $y = 5 + 2(x - 2)$ and $y = 7 + 2(x - 3)$ also express this same line. (Try it out!)

WHOLE-CLASS PRESENTATION

Students see the effects of changing each of the values x_1, y_1, and b. They see how changing these values in a linear equation in point-slope form changes the location of the line on the graph. Use the sketch **Point Slope2 Present.gsp.** Press the button *Show Point on Line*, and drag x_p to show how the calculated *y* value changes for different values of *x*. Then press the button *Show Function and Line*, and drag x_p again. Use the sliders to change x_1, y_1, and b and let students make observations. Do Q1–Q7 together as a class.

The slope-intercept form of a line is great if you know one special point: the *y*-intercept. But what if the point you know is an everyday, ordinary point such as $(3, -2)$ or $(-7, -7)$? In this case it's usually most convenient to use the *point-slope form* of a line, which you'll study in this activity.

SKETCH AND INVESTIGATE

1. Open the sketch **Point Slope2.gsp.**

To adjust a slider, drag the point at its tip.

You'll see an equation in the point-slope form $y = y_1 + b(x - x_1)$, with numbers filled in for b, x_1, and y_1. Adjust the sliders for b, x_1, and y_1, and watch the equation change. There's no line yet, but you can graph one.

2. Choose **Graph | Plot New Function.**

The New Function dialog box appears.

To enter b, x_1, and y_1, click their measurements in the sketch. To enter x, click the x key on the keypad.

3. Enter $y_1 + b(x - x_1)$ and click **OK.**

Sketchpad plots the function for the current values of the parameters b, x_1, and y_1.

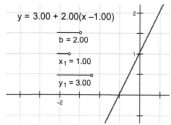

4. Select the new line and choose **Display | Trace Function Plot.**

If at any point you wish to erase traces left by the line, choose **Display | Erase Traces.**

Q1 Adjust slider b. You'll see that the line rotates around a single point. Change the values of x_1 and y_1; then adjust b again, focusing on where this point appears to be. What are the point's coordinates? How do they relate to x_1 and y_1?

To deselect all objects, click the **Arrow** *tool in empty space.*

5. Deselect all objects. Now select, in order, measurement x_1 and measurement y_1. Choose **Graph | Plot as (x, y)** to plot the point (x_1, y_1).

Q2 Adjust slider b again and observe what happens. Does this support your answer from Q1?

Q3 Describe the family of lines that forms when you change b.

Q4 Adjust sliders x_1 and y_1, one at a time. How would you describe the families of lines that form when varying each of these values? How do they compare to each other?

Q5 Summarize the roles that x_1 and y_1 play in the equation $y = y_1 + b(x - x_1)$.

Q6 Suppose you know that the slope of a line is 2 and that it contains the point $(1, 3)$. What is the equation in point-slope form for this line? Check your answer by adjusting the sliders in the sketch.

Q7 Write an equation in point-slope form for each of the lines described. When you finish, check each of your answers by adjusting sliders b, x_1, and y_1 so that the line is drawn on the screen.

    a. slope is 2; contains the point $(-2, 1)$

    b. slope is -1; contains the point $(-2, 1)$

    c. is parallel to the line $y = 4 + 3(x - 2)$; goes through the origin

    d. slope is $\frac{4}{5}$; x-intercept is $(2, 0)$

    e. contains the points $(2, 3)$ and $(-1, 4)$

    f. contains the points $(-3, 5)$ and $(4, 5)$

EXPLORE MORE

Q8 Try to construct a line through the points $(2, 3)$ and $(2, -2)$ by adjusting the sliders in the sketch. Explain why this is impossible and why this equation cannot be written in point-slope form.

Q9 Is it possible to construct the same line with different slider configurations? If not, explain why. If so, provide two different equations for the same line.

After completing this and the other two "Form of a Line" activities, students should have a good understanding of the three forms and the information each provides. Point out that all three forms can describe a single line.

Start with a discussion on why there are so many forms. Begin with "Lines are very important. Mathematicians have developed different ways of describing them depending on the information they have and the information they need."

End this activity by asking students to compare and contrast three equivalent equations, such as $y = (2/3)x + 2$, $2x - 3y = -6$, and $y = (2/3)(x - 3) + 4$. Have students convert equations between forms and decide which form is most appropriate for a particular purpose.

SKETCH AND INVESTIGATE

Q1 Slider a affects only the x-intercept. (Students may also notice that positive values of a correspond to positive x-intercepts and negative values to negative x-intercepts. The relationship is inverse, so no matter how large a becomes, the x-intercept never reaches 0.)

Slider b affects only the y-intercept. (The relationship between b and the y-intercept is similar to the relationship between a and the x-intercept.)

Slider c changes only the position of the line. (It translates the line parallel to itself.) It has no effect on the slope of the line.

Q2

a	2	3	12	-1	-3	0
x-intercept	3	2	0.5	-6	-2	undefined

Q3 The product of a and the x-intercept is c. This holds unless $a = 0$, in which case there is no x-intercept.

Q4

b	2	3	12	-1	-3	0
y-intercept	6	4	1	-12	-4	undefined

Q5 The product of b and the y-intercept is c. This holds unless $b = 0$, in which case there is no y-intercept.

Q6 For all values of c, the intercepts are in the ratio b:a. Students may also note that when $c = ab$, the x-intercept is b and the y-intercept is a.

Q7 x-int $= c/a$; y-int $= c/b$; slope $= -(a/b)$

Exploring Algebra 1 with The Geometer's Sketchpad
© 2012 Key Curriculum Press

Q8 a. $3x - 4y = 3$ b. $-2x + 5y = 10$ c. $x + 2y = 4$

EXPLORE MORE

Q9 To check the formulas for the slope and y-intercept, convert the standard form to the slope-intercept form:

$$ax + by = c$$

$$by = -ax + c$$

$$y = -\frac{a}{b}x + \frac{c}{b}$$

Thus the slope is $-a/b$ and the y-intercept is c/b. To check the formula for the x-intercept, substitute 0 for y:

$$ax + by = c$$

$$ax + b(0) = c$$

$$ax = c$$

$$x = \frac{c}{a}$$

Q10 The vector remains perpendicular to the line because the vector's slope is b/a, and the slope of the line is $-a/b$. These quantities are negative reciprocals, so the vector and line are perpendicular.

Q11 a. $5x - 3y = 0$ b. $-2x + 4y = 2$

The Standard Form of a Line

 Presenter Notes

In this presentation students will explore the graph of a line when the equation is in standard form ($ax + by = c$) and will see the effects of changing coeffecients.

1. Open **Standard Form Present.gsp.**

2. Introduce the activity and point out the equation displayed in standard form.

Q1 Ask how this equation looks different from linear equations expressed in point-slope and slope-intercept form. Here are some possible answers: x and y appear on the same side; y has a coefficient; and the slope does not appear explicitly in this equation.

Q2 Ask what the equation of this line would be in the other two forms. In slope-intercept form it would be $y = (-2/3)x + 2$. Multiple answers are possible in point-slope form. An answer using the point $(3, 0)$ is $y = (-2/3)(x - 3) + 0$, though infinitely many answers are possible using other points on the line.

3. Press the *Show Point on Line* button. Drag point P and observe the equation at the bottom. Explore how various combinations of x and y make the left side of the equation equal to the constant c on the right side.

4. Use the sliders to change a, b, and c, and explore the effects of these changes on the graph.

Q3 On page 2, complete the table by adjusting slider a to each specified value and double-clicking the table to record each value. Ask students to describe the relationship between the x-intercept and the values of a and c. (The x-intercept is equal to c/a.)

Q4 On page 3, complete the table by adjusting slider b, and ask students to describe the relationship between the y-intercept and the values of b and c. (The y-intercept is equal to c/b.)

Q5 On page 4, adjust slider c and ask students to describe its effect on the slope. (There is no effect.) Then adjust a and b to various positions, using whole numbers, and record the results by double-clicking the table. Ask students to describe the relationship between the two intercepts and the slope. (The slope is equal to $-a/b$.)

Q6 On page 5, press the buttons to show the various problems. For each problem, have students write down the answer, and then check their answers by manipulating the sliders.

Finish the presentation with a class discussion and summary of the relationship between the standard form coefficients and the characteristics of the line.

Exploring Algebra 1 with The Geometer's Sketchpad
© 2012 Key Curriculum Press

The Standard Form of a Line

When he's in the form of Clark Kent, he's a mild-mannered reporter who helps expose government corruption; in the form of Superman, he's a superhero who can get your cat out of a tree or save Earth from obliteration. Same person, but different forms. Similarly, the equation of the same line can appear in different forms. And like Clark Kent/Superman, the different forms of the equation are useful in different ways.

You are familiar with linear equations in slope-intercept form $y = mx + b$ and point-slope form $y = m(x - h) + k$. In this activity you'll explore linear equations written in standard form $ax + by = c$. At first, this form may not seem to convey much useful information. But as you'll soon see, understanding this form can lead to a whole new way of looking at lines.

SKETCH AND INVESTIGATE

1. Open **Standard Form.gsp.**

Q1 Adjust the sliders *a*, *b*, and *c* (by dragging the point at the tip), and describe the effect each of them has on the line.

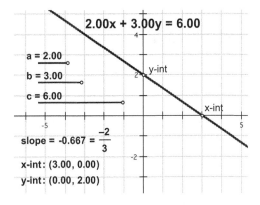

Select the both points and choose **Measure | Coordinates.**

2. Measure the coordinates of the *x*- and *y*-intercepts.

Q2 With slider *b* at 2.00 and slider *c* at 6.00, drag slider *a*. In a table like the one here, fill in the *x*-intercept for the given values of *a*.

a	2	3	12	−1	−3	0
x-intercept						

In this activity, the precision of measurements has been set to two decimal places. Be aware that coordinate and slope measurements may have been rounded off.

Q3 What is the relationship between the *x*-intercept and the values of *a* and *c*?

3. Make *c* = 12.00. Adjust slider *b* and observe the effect this has on the line.

Q4 In a table like the one here, fill in the *y*-intercept for the given values of *b*.

b	2	3	12	−1	−3	0
y-intercept						

Q5 What is the relationship between the *y*-intercept and the values of *b* and *c*?

4. Press the *Initial Position* button to return the sketch to its original state. Now adjust slider c, observing the line and the x- and y-intercepts.

5. Make $a = 2.00$ and $b = -4.00$. Once again, adjust slider c and observe the behavior of the line and the axis intercepts.

Q6 How do the values of a and b relate to the x- and y-intercepts of the line?

Q7 Based on your observations, write formulas for the x-intercept, y-intercept, and slope, using a, b, and c.

Q8 Find the equations of the following lines in standard form. First find the answers using paper and pencil; then check your answers by adjusting the sliders in the sketch.

    a. parallel to $3x - 4y = 12$; x-intercept at $(1, 0)$

    b. x-intercept at $(-5, 0)$; y-intercept at $(0, 2)$

    c. x-intercept at $(4, 0)$; containing the point $(2, 1)$

EXPLORE MORE

Q9 Check your formulas from Q7 by algebraically manipulating the equation $ax + by = c$.

A *vector* is a quantity with both a distance and a direction, such as *3 units to the right.* A vector represented as [2, 1] goes 2 units right and 1 unit up. Vectors are often represented by arrows.

Q10 Another property of a and b is this: Vector $[a, b]$ is perpendicular to the line. To see this, press the *Show Vector* button. The blue segment represents vector $[a, b]$. The *foot* of the vector has been arbitrarily placed at the origin, and the *head* is represented by the arrowhead. Adjust all three sliders, and notice that the vector remains perpendicular to the line. Explain why $[a, b]$ is perpendicular to the line.

Q11 Find the equations of the following lines in standard form. First find the answers using paper and pencil; then check your answers by adjusting the sliders in the sketch.

    a. perpendicular to the vector $[5, -3]$; passing through the origin

    b. perpendicular to the vector $[-2, 4]$; passing through $(5, 3)$

Exploring Algebra 1 with The Geometer's Sketchpad
© 2012 Key Curriculum Press

Lines of Fit

PROCESSING DATA

Q1 Answers will vary. The important thing is that students support their answers with arguments. This is a good question for discussion.

2. To show point labels as they plot the points, students can choose **Edit | Preferences | Text** and set Sketchpad to show labels automatically for all new points.

BEST-FIT LINE

The regression line constructed in this activity is not a least squares regression. Students only have to understand that their objective is to minimize the difference between estimated height and actual measured height.

Q2 In order to fit a line through all of the points, they would have to be collinear, and clearly, they are not. (In fact, it would not be possible to fit any function graph to the points, because two of them have the same x value.) However, it *is* possible to draw a line that comes close to all the points.

Q3 Answers will vary. Example: $y = 2.9x + 103$

Q4 The length of a segment represents the amount by which a data point misses the line vertically. It is blue if the point is above the line and red if it is below. The measurements are the lengths of the line segments in coordinate units.

Q5 This should be about the same as the answer to Q3, but this second method is more reliable.

Q6 About 159 cm.

EXPLORE MORE

With students at different stages of physical development, this experiment can generate an interesting range of data. You could also use this data to take a closer look at the question in Q1. Create separate lines of fit for girls and boys, and see if there is a significant difference. (Keep in mind that some students may be sensitive about having their body measurements taken. Modify the directions as needed.)

This problem involves an archaeologist who has uncovered a man's shoe, which would fit a foot of length 19 cm. She would like to get a rough estimate of the man's height. There are eight men and five women working at the dig site. She measures the foot length and the height of each man, but she does not measure the women.

1. Open **Lines of Fit Present.gsp.** Page 1 has a set of coordinate axes at an appropriate scale.

2. Select the table and choose **Graph | Plot Table Data.** This shows a scatter plot representing the foot lengths and heights of eight men.

3. Choose the **Line** tool to draw a line through the data plot. You must click in two places to define the line. Be careful not to click on any existing objects. You want both points to be independent points.

Normally, for such a rough estimate, you would simply draw the line so that it looks right. Adjust the line by dragging the two line points. When the class concurs on its position, find the equation and get an estimate.

4. Select the line. Choose **Measure | Equation.** Have students substitute 19 for x in the equation and solve for y.

5. How good was the line? Go to page 2. This has the same scatter plot. From each point, a vertical line segment represents the residual.

Q1 Ask what the line segments represent. For any foot length (x), the line estimates a man's height (y). Each line segment represents the difference between the actual height of a man and the estimate.

Q2 On the left side of the screen is a measurement labeled *total error.* This is the sum of the absolute values of the residuals. Tell students that this number represents the sum of the lengths of all of the colored line segments. Ask them how to use that number to improve the best-fit line. Help them understand that this number should be as small as possible.

6. Keeping an eye on the sum calculation, alternate between adjusting the two control points of the line to minimize the sum. Use the equation of the new line to make a new estimate of the man's height.

Encourage a discussion of the methods used. Was it right to exclude women from the sample? Should foot length and height have a linear relationship? How could the archaeologist improve the estimate?

It would not take long to conduct the similar measurements on the students.

In science you often have to gather data from observations and use them to make a reasonable guess about something you cannot see. If you have a lot of data and they show a consistent pattern, then you can have more confidence in your estimate.

PROCESSING DATA

Suppose you are an archaeologist, and your team has uncovered the remains of a shoe. The foot that fit the shoe would be 19 cm long, and by the type of footwear, you feel certain that it was a man's. In order to form a more complete picture of the shoe's owner, you would like to estimate his height. There are eight men working at the dig site. You decide to use them as your sample. You measure the foot length and height of each man. Here are your data.

Foot length (cm)	25	31	16	21	21	27	28	24
Height (cm)	189	195	149	174	158	169	180	172

Q1 There are also five women working at the site, but you do not measure them. Should you have included them in the sample? Explain.

1. Open **Lines of Fit.gsp.** Notice that the horizontal and vertical scales of the coordinate axes are different.

To label all the points, choose **Edit | Select All** and then choose **Display | Label Points.**

2. Choose **Graph | Plot Points.** Enter 25 in the first box and 189 in the second, then click **Plot.** Repeat this for each of the ordered pairs. When you have plotted the last point, click **Done.** Label all the points.

FINDING A LINE OF FIT

Q2 The simplest way (but not always the best way) to model the relationship between two variables is with a line. Is it possible to draw a line through the eight points on your screen?

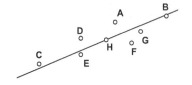

3. Choose the **Line** tool. Click in two places on the screen. When you do this, be careful not to click any of the existing points or axes.

4. Choose the **Arrow** tool. By dragging the two points that define the line, move it into a position where you think it best fits the data points.

This line represents your best guess. If you were to plot the foot length and height of another man, you would expect it to fall somewhere near the line.

5. Select the line. Choose **Measure | Equation.**

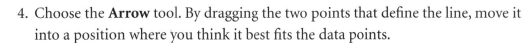

Q3 What is the equation of your line?

6. Press and hold the **Custom** tool icon, and choose |**Residual**| from the menu that appears. Click one of the data points and then click the line. The tool plots a line segment and a measurement. Repeat this at each data point.

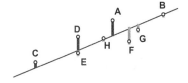

Q4 What is the meaning of the length and color of the line segments that are now attached to the data points? How are the measurements related to the line segments?

Perhaps you can adjust your line to fit the data better. The measurements are the magnitudes of the residuals. They tell you by how much each point misses the line vertically. In order to fit the data well, you need to minimize these values. The trouble is, if you move the line to make one smaller, you are probably making another one bigger.

7. Choose **Number | Calculate.** Compute the sum of the magnitudes of the residuals. This is the total amount by which the line misses the data.

Q5 Adjust the line to minimize that last calculation. What is the new equation for the line? Compare this with your answer to Q3.

8. Construct a point on the line. Measure its coordinates.

Q6 Drag the new point along the line until the *x*-coordinate is about 19. Based on your analysis, what was the approximate height of the man who owned the shoe?

EXPLORE MORE

Using classmates as a data sample, do some measurements and find a line of fit. There are many measurements that you could use—for example, hand span, arm span, and shoulder width.

7

Quadratic Equations

Modeling with Quadratic Equations: Where Are the Giant Ants?

Students manipulate a dynamic model of two similar polygons and explore issues of scale to better understand quadratic and linear relationships.

Graphing Quadratic Functions

Students plot the graph of a general quadratic function in standard form $y = ax^2 + bx = c$ and study the effects of changing the parameters a, b, and c.

Factoring Trinomials

Students factor trinomials using algebra tiles, explore why some trinomials can be factored and others cannot, and investigate the relationship between the factors and the coefficients of the trinomial.

Graphing Factored Quadratics

Students graph a function in the factored form $f(x) = a(x - r_1)(x - r_2)$ using the three parameters. They then investigate the relationships between the parameters and the graph.

Modeling with Quadratic Equations: Where Are the Giant Ants?

This activity introduces students to quadratic relationships and contrasts quadratic with linear relationships both numerically and graphically. Comparing the sizes of similar shapes will give students insight into one of the core ideas in the form and growth of animals.

Students are almost always surprised that length and area (or volume) do not grow in the same way. In order to confront their beliefs about growth, students should be encouraged to make and express their predictions prior to verifying them with Sketchpad.

Begin the discussion with a hypothetical question: Suppose ants could grow to 100 times their size (that is, 100 times as tall, 100 times as long, etc.). How would this affect their weight? How would it affect the area of one of their eyes? You can let students write down their guesses, then come back to the answer at the end of class. (The answer is that they would weigh 1,000,000 times as much, and the area of each of their eyes would be 10,000 times as large!)

This activity can provoke much class discussion and is well suited to classroom sharing. For instance, each student might choose to begin with a different type or shape of polygon. Students can also experiment with circles (see Explore More). By comparing their own results with those of classmates, they will gain a better appreciation of the quadratic growth of area for any shape. This will help them make conjectures regarding the growth of irregular shapes such as inkblots or amoebas.

Though the primary focus of this activity is length and area relationships, encourage students to investigate volume relationships. Once they have investigated area and talked about volume, they might also be able to predict what happens in higher dimensions.

SKETCH AND INVESTIGATE

6. To distinguish the new interior from the original, students can give it a different color by using the **Display | Color** submenu.

Q1 When $\overline{AB}$ is bigger, the scaled polygon is larger than the original and farther away from the dilation point. When $\overline{AB}$ is smaller, the scaled polygon is smaller and closer to the dilation point. When $AB = 0$, the image disappears.

7. Make sure students deselect all objects before selecting the two segments and measuring the ratio.

Q2 They are the same. (If they aren't, check that the sides chosen were really corresponding sides and that they were selected in the proper order—scaled polygon, then original polygon.)

8. Deselect all objects before selecting the interiors. To select an interior, click on it (and not on its sides or vertex points).

Q3 It is the same as the ratio of the corresponding sides, since perimeter is just the sum of the side lengths.

Q4 These ratios are all equal to the scale factor. Explanations will vary, but basically this is what a scale factor is: the ratio of each linear measurement in a scale drawing to that same measurement in the original.

COMPARING AREAS

Q5 Students should notice that this relationship is not linear. The ratio of the area grows faster than the ratio of the side lengths for scale factors larger than 1. To be more precise, the area ratio is the square of the side-length ratio. If the side-length ratio is 3 (3 : 1), the area ratio will be 9 (9 : 1). In terms of the "flat ant," if you double its waist size, you will quadruple the area of its skin.

10. Students may wish to hide the grid (by choosing **Graph | Hide Grid**), move the origin, and otherwise rearrange objects in the sketch to clean things up.

Q6 The graph makes sense because it appears to be a parabola (the graph of a quadratic function) and the relationship graphed is quadratic.

If students are too inexperienced with parabolas and quadratics to make this connection, they still may be able to see that when the shape is magnified, the area ratio grows faster than the side-length ratio as the scale factor gets bigger. This corresponds to the upward turn of the parabolic trace.

EXPLORE MORE

Q7 The same relationship holds with any two-dimensional shape. Students can compare circumferences or any pair of corresponding segments or distances in their figures to find that the ratio of linear measures is the

same as the scale factor. The ratio of the areas of the similar shapes is the square of the scale factor.

Q8 The ratio of the perimeters of the similar shapes is the same as the ratio of the side lengths, so the graph is $y = x$.

Q9 The ratio of the volumes of similar solids is the cube of the ratio of side lengths. This is a cubic function.

Q10 An animal that is twice as long, tall, and wide as another similarly shaped animal will have four times the surface area (skin) and eight times the volume. For this reason, relatively small evolutionary size increases place great demands on the overall system. The legs of the giant ant would never be able to support the mass of its body.

WHOLE-CLASS PRESENTATION

The goal of this presentation is to use a geometric model to let students explore issues of scale. They should come away with the sense of how a scale factor affects a drawing or a model. One-dimensional and two-dimensional quantities are affected differently in a scaling. One-dimensional quantities are magnified (or shrunk) by the scale factor, while two-dimensional quantities are magnified (or shrunk) by the square of the scale factor. This applies to all shapes—circles and irregular shapes as well as polygons.

Use **Quadratic Modeling Present.gsp** to explore with the class the effects of a dilation in two dimensions. The scale factor is controlled by the lengths of the two segments $\overline{AB}$ and $\overline{CD}$. Students should watch the ratios as you drag point B to vary the scale factor. Changing the scale factor to a few different integer values may help students see the relationship.

Page 2 shows three-dimensional shapes. Although Sketchpad cannot truly measure the volume of three-dimensional shapes, it can calculate the volumes as the product of base area and height. The table keeps track of the various ratios (of corresponding segments, corresponding areas, and volume) for different values of the scale factor. Drag point B to vary the scale factor. Try a few round numbers (integer values, 0.5, etc.). Double-click the table after each one to lock in the values. Let students think about these values and how they are related.

Finish by returning to the discussion of the giant ants and having students explain in their own words why ants cannot grow to even 10 cm in length, let alone to gigantic proportions worthy of monster status.

Modeling with Quadratic Equations:
Where Are the Giant Ants?

Some monster movies feature gigantic ants or other insects attacking and destroying cities. Did you ever wonder why such giant bugs don't exist outside the movies? Or why there aren't miniature elephants or elephant-sized ants? The answer to these questions actually relates to the evolution of species and to issues of scale. To better understand these issues, we'll scale down from three dimensions to two as you look at some "flat animals"—namely, polygons.

SKETCH AND INVESTIGATE

In this activity you'll explore what happens to a polygon's measurements as it gets bigger and smaller. So first you'll need to construct a polygon and its interior.

1. In a new sketch, use the **Segment** tool to draw a polygon with four, five, or six sides.

Each segment after the first should share an endpoint with the previous one. The last segment should connect back to the first. When you're done, you should have the same number of points as segments.

2. Click the **Arrow** tool in blank space to deselect all objects. Select all the points consecutively around the polygon, and choose **Construct | Interior.**

You should now have a polygon and its interior.

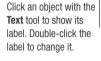
Click an object with the **Text** tool to show its label. Double-click the label to change it.

Next you'll make a scale copy of your interior. You'll use the ratio of the lengths of two segments as the scale factor. This will allow you to change the scale factor by dragging.

3. Construct segments AB and CD using the **Segment** tool.

4. Select in order $\overline{AB}$ and $\overline{CD}$ and choose **Measure | Ratio.** With the new ratio measurement still selected, choose **Transform | Mark Scale Factor.**

A   B
C   D
$$\frac{m\overline{AB}}{m\overline{CD}} = 1.37$$

5. Use the **Point** tool to draw a point outside the polygon. With the point still selected, choose **Transform | Mark Center.**

To select the entire polygon, start in empty space and draw a selection rectangle around it with the **Arrow** tool.

6. To construct the scale copy, select the entire polygon and choose **Transform | Dilate.** Click **Dilate** to dilate by the marked ratio.

$$\frac{m\,\overline{AB}}{m\,\overline{CD}} = 2.18$$

Dilation scales an object away from or toward a point. The scaled image appears. Drag point B to experiment with different scale factors.

Q1 What happens when $\overline{AB}$ is bigger than $\overline{CD}$? Equal? Smaller? What if $\overline{AB} = 0$? Experiment with dragging the center point and the vertices of your polygon.

7. Select a side of the scaled polygon and the corresponding side of the original polygon. Measure the ratio of these two segments.

Q2 Measure the ratio of a different pair of corresponding sides. What do you notice?

8. Measure the perimeters of the two polygon interiors by selecting both polygons and choosing **Measure | Perimeters.**

Choose **Number | Calculate** to open Sketchpad's Calculator. Click the perimeter measurements in the sketch to enter them into the calculation.

Q3 How do you think the ratio of these perimeters compares with the ratios you found in Q2? Why? Calculate the ratio to check your prediction.

Q4 How do the ratios of side lengths and perimeters compare with the scale factor you used for dilation? Why do you think that is?

COMPARING AREAS

You can think of perimeter as the flat ant's waist, and area as the surface area of its shell. When the ant's waist grows twice as big, what happens to the area of its shell? Think about this a moment before moving on.

Repeat step 8 and Q3, choosing **Area** instead of **Perimeter.**

9. Measure the areas of the two polygon interiors, and calculate the ratio of these two measurements.

Did you get the ratio you predicted? Drag point B until you've confirmed or refuted your prediction.

Q5 What have you discovered about the relationship between the ratio of lengths of similar figures and the ratio of their areas?

Often mathematical relationships become clearer when they are graphed. Next you'll plot the ratio of side lengths versus the ratio of areas for different scale factors.

10. Select in order the side-length ratio calculation and the area ratio calculation. Choose **Graph | Plot as (x, y).**

A coordinate system appears along with the plotted point. The coordinates of this point are (*side-length ratio, area ratio*) for the current scale factor.

Choose **Display | Erase Traces** if you wish to clear traces from the screen.

11. With the newly plotted point still selected, choose **Display | Trace Plotted Point.** Drag point *B* to experiment with different scale factors. Also drag the vertices of your polygon around—does anything change?

Q6 Explain why the shape of the graph makes sense given the relationship you discovered in Q4.

EXPLORE MORE

Q7 Your polygon probably didn't look much like a flat ant or elephant. For one thing, it was a little pointy. Repeat your investigation using a circle and its dilated image. Plot the ratio of radius measurements against the ratio of area measurements for the two circles. Does this change any of the relationships you found?

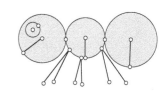

Select the ratios; then choose **Graph | Plot as (x, y).** Drag *B* to trace the shape of the graph.

Q8 What do you think the graph of perimeter ratio versus side-length ratio would look like? Make a prediction and then test it in Sketchpad.

Q9 Since animals are three-dimensional, let's move back to three dimensions. How does the ratio of the volumes of similar solids compare to the ratio of corresponding side lengths? Predict the answer, and then go to page 2 of **Quadratic Modeling.gsp** to investigate.

Q10 The introduction to this activity discussed the sizes of animals. How do you think your results (including the Explore More questions) help to explain why ants can't be the size of elephants? (*Hint:* Animals' masses are proportional to their volumes.)

Graphing Quadratic Functions

 ACTIVITY NOTES

TRACE A POINT

Q1 Point P has no maximum value. Its minimum value is -2.

Q2 The graph is a parabola. If students are not familiar with the correct term, they should still be able to describe its form as being roughly **U**-shaped.

GRAPH A FUNCTION

Q3 Changing parameter a stretches the graph vertically, but the y-intercept does not change.

When $a = 0$, this eliminates the first term from the quadratic function, so it becomes a linear function, $f(x) = bx + c$.

When $a > 0$, the curve opens upward. When $a < 0$, it opens downward.

Q4 When you change b, the graph retains its shape, but it is translated along a curved path. The y-intercept does not change.

When $b = 0$, the curve is symmetric with respect to the y-axis.

Q5 Changing parameter c translates the graph vertically.

When $c = 0$, the curve passes through the origin.

Q6 When you change a or b, the y-intercept does not change. On the y-axis, $x = 0$. Substitute 0 for x in the equation $y = ax^2 + bx + c$, and the result is $y = c$. Therefore it does not matter what values a and b have. The y-intercept is at $(0, c)$.

EXPLORE MORE

11. These values are used here in order to ensure that the function has real roots.

12. Students may have trouble with some of the more complicated calculations. Advise them not to delete anything if they get it wrong. They can double-click a calculation to edit it, even after they have plotted the point.

Q7 The points are, in order, the vertex, the y-intercept, and the two x-intercepts.

Graphing Quadratic Functions

In this activity students will observe the behavior of a plotted quadratic function in general form $f(a) = ax^2 + bx + c$ as you vary the three parameters a, b, and c. Students will form and evaluate conjectures about the effect of each of the parameters on the graph.

Use the prepared presentation sketch to eliminate some time-consuming parts of the student activity such as the entry of coordinate calculations.

1. Open **Graphing Quadratic Present.gsp.** Drag the sliders to show students how they control parameters a, b, and c.

The graph should appear after step 2. If it does not, the parameters may have moved it out of view.

2. Choose **Graph | Plot New Function.** The Calculator that appears has a key for the variable x. Enter the expression below. Enter the parameters by clicking them.

$$ax^2 + bx + c$$

Special properties appear when any parameter is zero. There are action buttons to make it easier to hit that value.

Q1 Tell students that you are going to change the value of parameter a gradually. Ask them what effect that will have on the graph. After some discussion, drag the a slider. Give special attention to the linear graph that appears when $a = 0$. Guide them to explain this by substituting zero for a in the function.

Q2 Challenge students to predict the effects of changing b. This is more difficult to predict or explain. Show the y-axis symmetry that appears when $b = 0$.

Q3 Show and discuss the changes that result from dragging the c slider. Again, stop briefly at zero so students can see that the curve goes through the origin.

Q4 There is one point that is always on the curve when parameters a or b are changed. If students did not notice, go back through those motions, and tell them to watch for it. The invariant point is the y-intercept. Help them explain this by substituting zero for x in the function. The result is c, no matter what the values of the other parameters are.

Q5 Plot each of the points below. In each case, challenge students to predict the location of the plotted point. To save time with the calculations, press the *Show Coordinates* button. For each point, select the coordinates in order and choose **Graph | Plot as (x, y).**

$$\left(-\frac{b}{2a}, c - \frac{b^2}{4a}\right); \ (0, c); \ \left(\frac{-b - \sqrt{b^2 - 4ac}}{2a}, 0\right); \ \left(\frac{-b + \sqrt{b^2 - 4ac}}{2a}, 0\right)$$

Graphing Quadratic Functions

A quadratic function is a polynomial function of degree 2. This is its general form:

$$f(x) = ax^2 + bx + c, \text{ where } a \neq 0$$

Quadratic functions are not as simple as linear functions, but they do have certain predictable properties, one of which is the shape of their graphs.

TRACE A POINT

1. In a new sketch, choose **Graph | Define Coordinate System.**

2. Construct a new point on the *x*-axis. Label it *x*.

To measure the *x*-coordinate, select point *x* and choose **Measure | Abscissa (x).** Double-click the **Text** tool on the measurement to change its label.

3. Measure the point's *x*-coordinate. Change the measurement label to *x*.

4. Choose **Number | Calculate** and calculate $x^2 - 4x + 2$. To enter *x*, click the *x* measurement in the sketch.

5. Drag point *x* left and right. Observe the value of the calculation.

6. Select in order the measurement *x* and the calculation of $x^2 - 4x + 2$. Choose **Graph | Plot as (x, y).** Label the new point *P*.

Q1 Drag point *x* and watch how the height of point *P* changes. What is the maximum height of point *P*? What is its minimum height?

To trace point *P*, select it and choose **Display | Trace Plotted Point.**

7. Turn on tracing for point *P*, and trace the path of *P* as you drag *x*.

This is what graphing is all about. With numbers, you can see the value of *y* for only one particular value of *x*. With a graph, you can see the values of *y* for a whole set of *x* values.

Q2 This is the graph of $y = x^2 - 4x + 2$. Describe the shape of this graph.

GRAPH A FUNCTION

In the equation $y = ax^2 + bx + c$, the right side is a function of *x*. With Sketchpad, you can define a function $f(x) = ax^2 + bx + c$ and graph the equation $y = f(x)$.

8. Choose **File | Document Options | Add Page | Blank Page.**

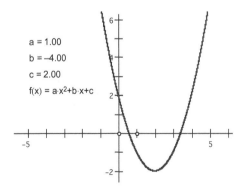

9. On the new blank page, define a coordinate system. Choose **Number | New Parameter** and label the new parameter *a*. Create two more parameters and label them *b* and *c*. You will change the values of the parameters later.

10. Choose **Graph | Plot New Function.** The Calculator dialog box that appears has a key labeled *x*. Enter $ax^2 + bx + c$ as the function definition. Click the parameters in the sketch to enter them.

This is a more convenient way of representing a graph in Sketchpad. You can change the parameters and the graph will change.

To make a parameter change more gradually, select it and choose **Edit | Properties | Parameter.** Change the Keyboard (+/−) Adjustments to 0.1.

Q3 Select parameter *a*. Change its value by pressing the **+** and **−** keys. What happens to the graph as you change parameter *a*? What is the shape of the graph when $a = 0$? Explain why. How does the sign of *a* influence the shape of the graph?

Q4 What happens to the graph as you change parameter *b*? What special property does the graph have when $b = 0$?

Q5 What happens to the graph as you change parameter *c*? What special property does the graph have when $c = 0$?

Q6 When you change *a* or *b*, is there always one point that does not move? What point is that? Explain why it remains fixed when all of the other points are moving.

EXPLORE MORE

Below are the coordinates of four special points on the graph. Follow the instructions to plot them on the graph.

$$\left(-\frac{b}{2a}, c - \frac{b^2}{4a}\right); \ (0, c); \ \left(\frac{-b - \sqrt{b^2 - 4ac}}{2a}, 0\right); \ \left(\frac{-b + \sqrt{b^2 - 4ac}}{2a}, 0\right)$$

11. Set the parameters back to the values they had in the first section:

$$a = 1, b = -4, c = 2$$

12. Choose **Number | Calculate** and calculate the *x*-coordinate of the first point. Repeat for the remaining seven *x*- and *y*-coordinates.

13. Select a coordinate pair in order and choose **Graph | Plot as (x, y).** Repeat this for each of the points.

Q7 Describe the significance of each point. Vary the parameters to make sure your descriptions are accurate no matter what the shape of the graph is.

Exploring Algebra 1 with The Geometer's Sketchpad
© 2012 Key Curriculum Press

Factoring Trinomials

The advantage of Sketchpad's algebra tiles is that they can be attached to each other and are based on variables rather than being of fixed size. This allows students to drag the sliders for x and y to see that x and y are variables and that the relationships work no matter what their values.

SKETCH AND INVESTIGATE

Q1 There are six tiles of three types, with each term's coefficient determining the number of tiles of that type.

Q2 The length is $(x + 2)$ and the width is $(x + 1)$. When you multiply these, you get the original trinomial $x^2 + 3x + 2$.

Q3 The factors of $x^2 + 3x + 2$ are $(x + 2)$ and $(x + 1)$.

Q4 Answers will vary depending on the value of x.

Q5 $3x + 6 = 3(x + 2)$

$x^2 + 7x + 12 = (x + 4)(x + 3)$

$2y^2 + 5y + 2 = (y + 2)(2y + 1)$

$y^2 + 3xy + 2x^2 = (y + 2x)(y + x)$

Q6 The sides of the rectangle are $(x + 2)$ and $(x + 3)$. Some students will have $(x + 2)$ as the length and $(x + 3)$ as the width, and others will have it vice versa.

Q7 You'll need four x tiles.

Q8 The factors of $x^2 + 4x + 3$ are $(x + 3)$ and $(x + 1)$.

Q9 When you make x smaller, the tiles become smaller and spaces appear between them.

Q10 You can drag the tiles back into a rectangle. The factors remain the same: $(x + 3)$ and $(x + 1)$.

Q11 The factors of $x^2 + 5x + 4$ are $(x + 4)$ and $(x + 1)$.

Q12 When you drag the slider to change the value of x, the tiles get larger or smaller, but they remain attached in the form of a rectangle.

Q13 a. $(x + 6)(x + 1)$ b. $(y + 4)(y + 2)$
c. $(y + 6)(y + 2)$ d. $(x + 2y)(x + 2y)$
e. $(2x + 1)(x + 1)$ f. $(4y + 3)(y + 1)$

In all cases, the two factors can be reversed.

Q14 You cannot make a perfect rectangle with this group of tiles. The x's can all be lined up along one side of x^2, or three along one side and one along the other, or two along either side. These are the only possibilities, and in none of these cases do the six unit squares fit in.

Q15 The second term is the sum and the third term is the product of the same two numbers. If the factors are $(x + 4)$ and $(x + 2)$, the coefficient of the second term is the sum of 4 and 2, and the third term is the product of 4 and 2. Thus

$$x^2 + 6x + 8 = (x + 4)(x + 2)$$

EXPLORE MORE

Q16 You factor this binomial the same way, by making the tiles into a rectangular shape.

The factors are x and $(x + 3)$.

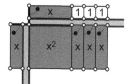

Q17 The only expressions that can be represented by more than one differently shaped rectangle have a constant factor common to each term. (This activity contains no such expressions.) An example is $2x^2 + 12x + 16$. This expression can be factored and represented with algebra tiles as either $(2x + 8)(x + 2)$ or $(2x + 4)(x + 4)$. Neither of these forms is fully factored. The full factorization is $2(x + 4)(x + 2)$. A model with algebra tiles would consist of two separate, identically sized rectangles.

In this presentation students will see how algebra tiles model factoring of trinomials, will see why some trinomials cannot be factored, and will see that the relationships explored hold even when the values of the variables are changed.

1. Open **Factoring Trinomials.gsp.** Ask students how the tiles correspond to the coefficients of the trinomial.

Q1 Press the *Arrange Tiles* button. After the rectangle appears, ask students to identify the length and width of the rectangle. Also ask them to identify the parts of the rectangle that correspond to the terms of the original trinomial.

Q2 Ask students to write the area in two different ways. Length times width gives $(x + 2)(x + 1)$, and counting the shapes gives $x^2 + 3x + 2$.

2. Use the Calculator to make sure that both ways have the same numeric result.

> Choose **Number |
> Calculate** to make
> the Calculator appear.
> To enter x into your
> calculation, click
> the value of x in the
> sketch.

3. Drag the x (blue) slider. The rectangle comes apart. Press the *Arrange Tiles* button to put it back together. On page 4 you'll make one that stays together.

4. On page 2, ask students to give the unfactored and factored expressions for the area of each rectangle.

5. On page 3, drag the x^2 tile to an empty area of the sketch.

6. Move one of the x tiles down so it's near the x^2 tile. Drag the black point of the x tile to show how you can orient it vertically or horizontally. Leave it vertical and drag the tile to align it on the right side of the x^2 tile.

Q3 Ask students how to drag each of the remaining tiles to form a rectangle. Have them summarize in their own words the factorization of this trinomial.

7. On page 4, choose the **x^2** custom tool and click once to create an x^2 tile.

8. Choose the **x** tool and click the upper-right corner of the square. Use the **Arrow** tool to make the x tile vertical.

9. Use the **x** tool and the **unit** tool to complete the rectangle.

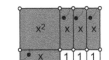

Q4 Ask, "What is the factorization of $x^2 + 4x + 3$?"

Q5 Have students volunteer to use the tools to factor the problems on pages 5–10.

Q6 On page 11, have students write trinomial and factored form for each rectangle.

Q7 On page 12, try to factor $x^2 + 4x + 6$. What goes wrong?

Finish with a class discussion on factoring and on how the algebra tiles help illustrate the equivalence of the two forms.

Factoring Trinomials

In this activity you'll factor trinomials visually by using Sketchpad algebra tiles.

SKETCH AND INVESTIGATE

1. Open **Factoring Trinomials.gsp.**

Q1 How many tiles are there for the trinomial $x^2 + 3x + 2$? How many different kinds? How do these tiles represent this particular trinomial?

2. To factor the trinomial, you must arrange its tiles into a rectangle. Press the *Arrange Tiles* button to do so.

Q2 What is the length of the resulting rectangle in terms of x? What is the width? Multiply these two binomials. What result do you get?

Q3 What are the factors of the trinomial $x^2 + 3x + 2$?

Use Sketchpad's Calculator to calculate the two results. With the Calculator open you can click the value of *x* in the sketch to enter it into your calculation.

Q4 Calculate the numeric result both ways to make sure the trinomial really is equal to the product of its factors. Record the value of x and the two results.

Q5 Page 2 contains four rectangles made from tiles. Write the area of each rectangle in trinomial and factored form. Drag the sliders to make sure the relationships hold for different values of the x and y variables.

On page 3 is another trinomial to factor. You will make this rectangle yourself.

3. Drag the x^2 tile to an empty area of the sketch.

4. Move one of the x tiles so it's near the x^2 tile. Then drag the black point of the x tile to see how you can orient it either vertically or horizontally. Leave it vertical.

5. Drag the x tile so it's aligned on the right side of the x^2 tile.

6. Drag the remaining tiles so they are aligned with the tiles you have already placed and so all the tiles form a rectangle. You may need several tries to make this work.

Q6 What are the length and width of the rectangle? What are the factors?

7. Move your rectangle to the upper-left corner of the frame. Then use the custom tools to make trains outside the frame that represent the factors of the trinomial.

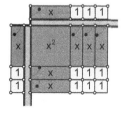

To use the **x^2** custom tool, press and hold the **Custom** tool icon and choose **x^2** from the menu that appears.

8. Page 4 has another trinomial, but no tiles. Choose the **x^2** custom tool and click once in the sketch to create an x^2 tile.

Exploring Algebra 1 with The Geometer's Sketchpad
© 2012 Key Curriculum Press

Q7 How many *x* tiles will you need? Use the **x** custom tool to make them.

9. Use the **1** tool to make as many unit tiles as you need, and then arrange the tiles into a rectangle.

Q8 What are the factors of this trinomial?

Even if you change the value of *x*, the tiles should still fit into a rectangle.

Q9 Make *x* smaller by dragging the blue slider. What happens to the tiles?

Q10 Can you drag the tiles back into a rectangle? What are the factors now? Did changing the value of *x* change the factors?

On the next page you will attach the tiles so that they don't come apart.

10. On page 5 is another trinomial. Use the **x^2** tool to create the first tile.

11. Attach an *x* tile to the right side of the x^2 tile by clicking the **x** custom tool on the upper-right vertex of the x^2 tile. Use the **Arrow** tool to drag the black point so that the *x* tile is oriented vertically.

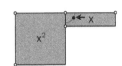

12. Attach an *x* tile to the bottom of the x^2 tile by clicking the **x** tool on the lower-left vertex of the x^2 tile. Use the **Arrow** tool to orient this new tile horizontally.

If you attach a tile in the wrong place, use **Edit | Undo** and then try again.

Q11 Construct the remaining tiles, attaching each new tile to the ones you have already placed. What are the factors of this trinomial?

Q12 Drag the *x* slider to change the value of *x*. What happens to the tiles?

Q13 Build and factor the following expressions on the remaining pages of the sketch. Draw the models on your paper.

    a. $x^2 + 7x + 6$ b. $y^2 + 6y + 8$ c. $y^2 + 8y + 12$

    d. $x^2 + 4xy + 4y^2$ e. $2x^2 + 3x + 1$ f. $4y^2 + 7y + 3$

Q14 On page 12, try to factor $x^2 + 4x + 6$. Describe the problem you encounter.

Q15 Describe how the second and third coefficients of a trinomial are related to the factors when the leading coefficient (the coefficient of the x^2 term) is 1.

EXPLORE MORE

Q16 Use the algebra tiles to factor the binomial $x^2 + 3x$.

Q17 Is there more than one rectangular shape that can be made to model any of the factorable expressions in this activity? See if you can come up with a factorable expression for which more than one rectangle can be made.

Graphing Factored Quadratics

 ACTIVITY NOTES

SKETCH AND INVESTIGATE

Q1 The x-intercepts are -1 and 4, the same as the roots. If you substitute either of these numbers for x in the function definition, one of the factors will be zero, so the corresponding point must be on the x-axis.

Q2 Changing parameter a stretches the graph vertically from the x-axis while the x-intercepts remain constant. When $a > 0$, the parabola opens upward. When $a < 0$, it opens downward. When $a = 0$, the graph coincides with the x-axis. This is because the function definition becomes $f(x) = 0$.

Q3 When you change the roots, the x-intercepts change accordingly. When the roots are equal, there is only one intercept, and the parabola is tangent to the x-axis at that point.

Q4 If you know the roots, that gives you parameters r_1 and r_2, but you still need to know a in order to complete the function definition.

EXPLORE MORE

This last section guides students through some calculations and constructions. They must then finish with less guidance.

Q5 The coordinates of the x-intercepts are $(r_1, 0)$ and $(r_2, 0)$. The y-intercept is at $(0, f(0))$, or $(0, ar_1r_2)$.

Q6 In order to write a quadratic function in factored form, with real numbers, it must have roots. If the graph does not intersect the x-axis, this method will not work. One simple example is the function $f(x) = x^2 + 1$.

Exploring Algebra 1 with The Geometer's Sketchpad

In this presentation students will observe the graph of a quadratic equation in factored form:

$$f(x) = a(x - r_1)(x - r_2)$$

Students will make conjectures about the behavior of the graph and will confirm or modify those conjectures as the parameters a, r_1, and r_2 change. They will see how to find the axis of symmetry, vertex, and intercepts from this form of the equation.

1. Open **Factored Quadratic Present.gsp.** The three sliders control the parameters a, r_1, and r_2. Demonstrate how the sliders control the parameters. Tell the class that during the construction, they should be watching for the relationships between these three parameters and the graph.

For the x, use the x key on the Calculator keypad. To enter a parameter, click it in the sketch.

2. Choose **Graph | Plot New Function.** Enter $a(x - r_1)(x - r_2)$.

Q1 Ask the class for the values of the x-intercepts. [They are the roots.] Drag the root sliders one at a time so that students can see that each slider changes only one root. To prove this, substitute one of the roots into the function definition, and show that it becomes zero.

Q2 Ask students to predict what will happen when you change parameter a. If they need a hint, ask them if changing a will change the x-intercepts. Drag the a slider slowly to show the family of parabolas having the given roots. Spend a moment on the special case of $a = 0$.

Q3 Ask what will happen if the roots are equal. Press the *One Root* button to move r_1 to r_2. The curve will be tangent to the x-axis at the one root.

Q4 Ask for the coordinates of the x-intercepts. [They are $(r_1, 0)$ and $(r_2, 0)$.]

3. Choose **Number | New Parameter.** Change the label to *zero*, and make the value 0.

4. Select in order parameters r_1 and *zero*. Choose **Graph | Plot as (x, y).** Use the same method to plot $(r_2, 0)$.

5. Calculate and plot the y-intercept $(0, f(0))$ and vertex $((r_1 + r_2)/2, f((r_1 + r_2)/2))$. You can use the Calculator to compute $f(0)$ by clicking the function itself in the sketch and then entering the argument (0 in this example).

Q5 Compare this function with a quadratic function in general form $f(x) = ax^2 + bx + c$. What are the advantages of the different forms? Is it possible to convert one form to the other?

Graphing Factored Quadratics

When you use quadratic functions to model real-world situations, you often begin with the roots. The roots are the x values for which the function is equal to zero. For the flight of a ball, the roots could represent the horizontal locations where it left the ground and where it landed. In cases like this, you use a quadratic function in factored form: $f(x) = a(x - r_1)(x - r_2)$, where r_1 and r_2 are the two roots.

SKETCH AND INVESTIGATE

1. Open a new sketch. Choose **Graph | Define Coordinate System.**

The square brackets indicate a subscript. The labels *r[1]* and *r[2]* will appear as r_1 and r_2 in the sketch.

2. Choose **Number | New Parameter** and label the new parameter a. Create a second parameter, labeling it *r[1]* and giving it a value of -1. Create a third parameter *r[2]* with a value of 4.

3. Choose **Graph | Plot New Function.** The Calculator dialog box that appears has a key labeled x. Enter $a(x - r_1)(x - r_2)$. To enter a, r_1, or r_2, click the parameter in the sketch.

a = 1.00

$r_1 = -1.00$

$r_2 = 4.00$

$f(x) = a \cdot (x - r_1) \cdot (x - r_2)$

Q1 What are the x-intercepts of the graph? Explain why this is true.

To make a parameter change more gradually, select it and choose **Edit | Properties | Parameter.** Change the Keyboard (+/−) Adjustments to 0.1.

4. Select one of the parameters. Change it by pressing the + and − keys. Experiment with all three parameters—a, r_1, and r_2.

Q2 Describe what happens when you change parameter a. How does the sign of a influence the shape of the graph? What happens when $a = 0$?

Q3 What happens when you change r_1 or r_2? What special property does the graph have when the two roots are equal?

Q4 You may know the roots of a certain quadratic function (as in the ball example), but that information alone is not enough to derive the function and draw the graph. Explain why.

EXPLORE MORE

In this section you will use algebraic and geometric concepts together in order to plot other objects related to the graph.

The line of symmetry of this graph is vertical, and it must intersect the x-axis at a point midway between the two roots. Therefore, the x-coordinate of this point must be the mean of the two roots.

Exploring Algebra 1 with The Geometer's Sketchpad
© 2012 Key Curriculum Press

5. Calculate the mean by choosing **Number | Calculate** and entering $(r_1 + r_2)/2$.

6. Create another new parameter. Label it *zero*, and set its value equal to zero.

7. Plot the point $((r_1 + r_2)/2, 0)$. To do so, select in order the calculation $(r_1 + r_2)/2$ and the parameter *zero*, and choose **Graph | Plot as (x, y)**.

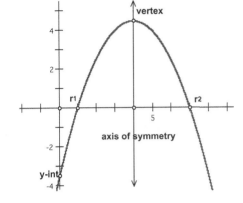

To construct the parallel, select the plotted point and the *y*-axis. Choose **Construct | Parallel Line.**

8. Construct the axis of symmetry by constructing a line parallel to the *y*-axis through the plotted point.

The vertex is on the axis of symmetry, so its *x*-coordinate must be $(r_1 + r_2)/2$. It is also on the parabola, so its *y*-coordinate is $f((r_1 + r_2)/2)$.

9. Calculate the *y*-coordinate of the vertex by choosing **Number | Calculate.** Click the function $f(x)$ in the sketch and then the calculation $(r_1 + r_2)/2$. Click **OK.**

10. Plot the point $((r_1 + r_2)/2, f((r_1 + r_2)/2))$. Label it *vertex*.

Q5 What are the coordinates of the two *x*-intercepts and the *y*-intercept? Plot these points and label them appropriately.

11. Change the values of all three parameters, and observe the behavior of the intercepts, axis of symmetry, and vertex.

Q6 You can still change the parameters to show graphs of other quadratic functions. Are there any quadratic functions that you cannot show using this sketch? Explain.